Animal Experimentation and the Future of Medical Research

Animal Experimentation and the Future of Medical Research

Proceedings of a Conference organized by the
Research Defence Society, held on 26 April 1991
at the Royal Society, London

Editor

Jack H. Botting

PORTLAND PRESS
London and
Chapel Hill

Published by Portland Press, 59 Portland Place,
London WIN 3AJ, U.K.
In North America orders should be sent to Portland Press Inc.,
P.O. Box 2191, Chapel Hill, NC 27515-2191, U.S.A.

© 1992 Portland Press, London

ISBN 1 855 78 038 0   **ISSN 0966-4068**

British Library Cataloguing in Publication Data
A catalogue record for this book is available from
the British Library

Typeset by Unicus Graphics Ltd, Horsham, Sussex
and Printed in Great Britain by Whitstable Litho Printers Ltd

Lord Adrian MD, FRCP, FRS (*Professor of Cell Physiology, University of Cambridge; Master of Pembroke College, The Master's Lodge, Pembroke College, Cambridge CB2 1RF, U.K.*)

Professor Sir Walter Bodmer FRS, FRCPath (*Director of Research, Imperial Cancer Research Fund; ICRF, Lincoln's Inn Fields, London WC2A 3PX, U.K.*)

Professor Sydney Brenner CH, DPhil, FRCP, FRS (*Fellow of King's College, Cambridge; Director Molecular Genetics Unit, University of Cambridge; Member of Scripps Research Institute, La Jolla, California; Molecular Genetics Unit, Addenbrooke's Hospital, Hills Rd, Cambridge CB2 2QH, U.K.*)

Professor David Hubel MD (*John Franklin Enders University Professor, Harvard Medical School; Nobel Prize for Physiology or Medicine, 1981; Harvard Medical School, 220 Longwood Avenue, Boston, MA 02115, U.S.A.*)

Professor David Peters FRCP (*Regius Professor of Physic, University of Cambridge; Fellow of Christ's College, Cambridge; School of Clinical Medicine, Addenbrooke's Hospital, Hills Road, Cambridge CB2 2QQ, U.K.*)

Doctor David Rees BSc, PhD, DSc, FRS, FRSC, FIBiol (*Secretary, Medical Research Council; MRC, 20 Park Crescent, London W1N 4AL, U.K.*)

Professor Sir John Vane BSc, DPhil, DSc, FRS (*Director, William Harvey Research Institute; Nobel Prize for Physiology or Medicine, 1982; William Harvey Research Institute, St Bartholomew's Hospital Medical College, Charterhouse Square, London EC1M 6BQ, U.K.*)

Contents

Protest against the use of animals to extend scientific knowledge has been rife since the experimental biological disciplines burgeoned in the 19th Century. Some of the early experiments, performed without the use of the refined anaesthetic techniques available today, may appear unsophisticated to modern eyes, yet the antivivisection protest at that time was orderly, reflecting the benevolence and gentility of the Victorian middle class. Paradoxically, today, when animal welfare is considered of paramount importance from the viewpoints of both humanity and good science, and discomfort of the animal under experiment is kept to a minimum, the protest has at times become destructive and violent to the point of deliberate endangerment of life.

One of the causes of the extreme actions taken by some animal rights groups is their mistaken association of animal experimentation with profit motive in the pharmaceutical industry (this of course appeals to the natural anarchic instincts of youth), coupled perhaps with the psychological need for an outlet for fantasy and excitement, which modern civilized society does not provide for all.

Much of the problem, however, results from factual misrepresentation, by the well-funded antivivisection societies, of the contribution that animal experiments make to advances in medical science. Some of the statements from those that oppose animal experiments, for example the suggestion that no medical advances have ever been achieved by animal experiments, are so manifestly untrue that, in the past, the majority of experimenters have regarded them as not worthy of attention. This attitude has encouraged antivivisection organizations to produce even more extreme statements, which through their audacity inevitably influence the lay public.

To counter this propaganda the Research Defence Society organized a one-day symposium, held at the Royal Society in London on 26 April 1991. The conference enabled biomedical scientists of unquestionable stature to give presentations describing the contribution of animal experimentation to past advances in their particular fields, as well as the contribution one could reasonably anticipate from similar techniques in the foreseeable future. During the lectures and ensuing discussion, emphasis was often placed upon advances that have made some types of animal experiments no longer necessary. For scientific and humane reasons such progress was received with unqualified approval. However, the undeniable fact that will be apparent to any unbiased individual who examines these proceedings is that, if we are to expect effective treatments for the remaining lethal infectious diseases, cancers, inherited diseases, traumatic nerve damage etc., some animal experimentation will continue to be essential.

As emphasized by the President of the Research Defence Society, Lord Adrian, in his introduction and reiterated by speakers throughout the meeting: "animal experiments should only be used if there is no other way to conduct the research in question and if all reasonable steps are taken to keep any distress or suffering to a minimum".

Jack H. Botting
Research Defence Society
1992

The ethics of animal research: introduction

Lord Adrian FRS
President, Research Defence Society

This conference is being held by the Research Defence Society, a society which has existed for more than 80 years as an important voice of the scientific and medical community in the debate about the use of animals in research.

The purpose of this meeting is to examine, from a scientific and medical perspective, the future role of animal experiments in medical research. We are extremely fortunate to have with us some of the most eminent clinicians and scientists in this country and from the United States to speak about their particular fields of work and to consider the importance of animal experiments in each case.

In this introduction, I want to take a slightly different perspective. Animal experimentation is of course a scientific method and like all science it exists in a social context. The most conspicuous aspect of that social context is the existence of organized opposition to animal experiments. For the past 150 years at least, animal research has been accompanied by protests, sometimes violent, from groups and individuals who consider that the use of animals in such research is unethical. Let me therefore begin by trying to examine, in a fairly elementary way, the ethical basis for animal research and to set the following scientific and clinical examination in an ethical context.

Let me start with what I would regard as a generally acceptable statement about the ethical justification for using animals in research.

> 'The use of living animals in scientific research can be considered justified if it is likely to produce appreciable benefit to society, if there is no other way to conduct the research in question and if all reasonable steps are taken to keep any distress or suffering to a minimum.'

The first point is that the fundamental justification for animal experimentation is the same as for any other scientific research: the discovery of new knowledge and, through this, the potential to benefit the society in which we live. The advancement of human civilization across the whole world has depended largely on the use of knowledge and technology to advance communication of ideas, to transport people and goods, to cure diseases and to enrich our lives in many other ways.

As an engine of change, scientific research has been fundamental in building modern civilization, and a striking example of that progress during the past century can be seen in our ability to diagnose and treat both human and animal diseases. In fact, the practice of medicine has changed beyond recognition in the past 200 years and to a large degree these changes have depended on experimental work in animals.

The fact that we use animals for the benefit of human society is quite undeniable. Human society does use animals in many ways: for food, for clothing, for companionship, for entertainment as well as for research.

However, there is a school of thought which regards many of these uses as morally objectionable. The philosophy of 'animal rights' holds that it is fundamentally wrong for humans, as a dominant species, to use other species for their own benefit, arguing that they should grant equal or near equal rights to other animals. These ideas, I think it has to be said, do not relate easily to the way in which most of us live our lives. Humans quite naturally accord the greatest priority to their own welfare. The animal rights philosophy may be held conscientiously by those who seek to abolish the use of animals in research and their exploitation in other ways by humans, but it cannot be held by those of us who believe that human welfare can justify animal use in research. For us it is a question of human duties rather than animal rights.

The second important point in my original statement is the caveat 'if there is no other way'. I think that most of us connected with animal experimentation consider that this is the most important consideration involved. As human beings, as moral agents, we recognize that we have a duty to safeguard the welfare of animals in our care. We use animals for human benefit but our sense of morality demands that this use should be combined with due regard for the welfare of those animals. The first principle of this regard for their welfare is that we should not use animals unnecessarily.

The most obvious way in which laboratory animals could be used unnecessarily is for research that is pointless, trivial or improperly designed so that it cannot produce any useful results. This point was emphasized back in the beginning of the 19th Century by those scientists, Marshall Hall particularly, who addressed this question. The protection that we have against this is to evaluate the scientific or clinical merit of proposals for research. In the case of animal research this is usually performed twice. In the normal course of events, applications for funding for research are subject to peer review in competition with other research proposals. If the research involves the use of animals it is then subject to a further review by the Home Office Inspectorate who are by law required specifically to assess the justification for the use of animals, weighing up the potential benefits of the research against the potential distress and suffering on the part of the animals.

The second caveat in the initial statement is that 'all reasonable steps are taken to keep any distress or suffering to a minimum'. In the past two decades we have seen very considerable improvements in the way that research animals are housed, cared for and used. Many will be aware that the Government has recently announced that £75 million has been allocated to help bring the standard of animal housing in our universities and polytechnics up to the most modern level—standards which are defined not just in terms of cage sizes but which include details of floor and wall surfacing, permitted humidity and temperature ranges and number of air changes per hour. It is important to emphasize that the universities and polytechnics use only 23% of the research animals in this country and that the £75 million is only the cost of upgrading—a fraction of the total investment in the very high standard of laboratory animal housing which is the norm in the U.K. The Home Office Inspectorate review experimental procedures proposed in every animal research project to ensure that only minimal numbers of animals are used and that any potential for suffering is kept to a minimum. This is not a passive function and I am advised that it is by no means unusual for the Inspectorate to require a number of changes to be made in a research proposal before the Home Office will grant the necessary licence to permit the research to be conducted.

The statutory Animals Procedure Committee also conducts reviews of particular animal research techniques and even entire areas of animal research to set bench-marks for the Inspectorate in their task of working towards those important minimums. It is with very good reason that the laws which control animal research in this country have been called the most comprehensive in the world. They were the product of long deliberation and negotiation between the Government, the scientific community, the veterinary profession, the medical profession and the responsible welfare animal community, most ably co-ordinated in the discussions in the House of Lords by Lord Houghton of Sowerby.

In the 1970s there was broad agreement that the old 1876 Act controlling animal research in the U.K. needed to be replaced. The Research Defence Society devoted a full decade of its effort to the negotiations leading up to the new legislation. Under my predecessor as its President, Lord Halsbury, the Society was responsible for the introduction of the Laboratory Animals Protection Bill into the House of Lords in 1979 and for initiating the House of Lords Select Committee inquiry into the whole subject of legislation for animal experiments. This has been acknowledged as having significantly advanced the legislative process which was completed some 7 years later with the passage of the Animals (Scientific Procedures) Act in 1986. It is that Act which presently defines the structure of the controls under which all animal experiments are carried out in this country.

The immediate purpose of this conference is to consider the future of animal research from a clinical and scientific perspective. However, no consideration of animal research is complete if it ignores the ethical context and justification for that research. I have attempted to sketch in outline that justification, but we will hear the details throughout the rest of the conference. The enormous advances in medical research which our very distinguished speakers will be describing are of the very essence of that ethical justification.

Animal experimentation and the future of medicine

D. K. Peters
Regius Professor of Physics, University of Cambridge

As a practising physician I am daily reminded of the advances that have taken place in medicine which have radically altered the shape of medical practice. I am asked to consider the future of medicine against a background of accelerating change in medical practice as the revolution in biology is applied to its most important field: the prevention, diagnosis and treatment of human disease. Although emphasis is placed on disease in man the new biology has major implications for the practice of veterinary medicine.

Futurology is hazardous and it is impossible to have a sense of direction without taking a historical perspective. In the course of this essay I shall therefore trace the history of certain areas of medical research to provide an impression of the dramatic impact of animal experimentation on the improvement of human health, to give indications of how the momentum of these discoveries may affect our future needs, and to indicate how modern approaches to animal experimentation—through for example genetic engineering—can provide powerful methods for advancing medical knowledge.

Figure 1 depicts a striking example of modern medicine: a new-born baby in an intensive care unit with all the paraphernalia associated with high technology medicine. The advances that have been responsible for the care of this baby have depended heavily on animal research. The baby has myasthenia gravis, a disorder in which transmission of nerve impulses between nerves and voluntary muscles is impaired. We now know that this impairment is usually due to autoimmunity, where antibodies are formed which block neuromuscular transmission. The physiological basis of neuromuscular transmission was unravelled by research in experimental animals. Likewise, the understanding of the physiology of the immune system had its roots in the study of immunity in experimental animals. Elucidation of the role of autoantibodies in myasthenia gravis also required the use of animal experimentation. The treatment of this patient, as can be seen in Figure 2, had a splendid and happy outcome. The mother, you will note, has a tracheostomy and would have been treated on a ventilator in order to sustain respiration. The knowledge of the physiology of the respiratory system in turn depended on experimentation in animals. This single example provides testimony to the pervasive influence of animal experimentation on clinical practice.

Turning from a single patient to a population health problem, may I remind you that high blood pressure, ischaemic heart disease and heart failure are the most common causes of mortality in the U.K.

The development of treatments of cardiovascular disease represents a triumph of pharmacological research. Most of the drugs now used to treat these conditions were not available when I was a medical student some 30 or so years ago. Consider the treatment of high blood pressure and heart failure by angiotensin-converting-enzyme (ACE) inhibitors. This approach arose from studies of experimental models of high blood pressure brought about

Figure I

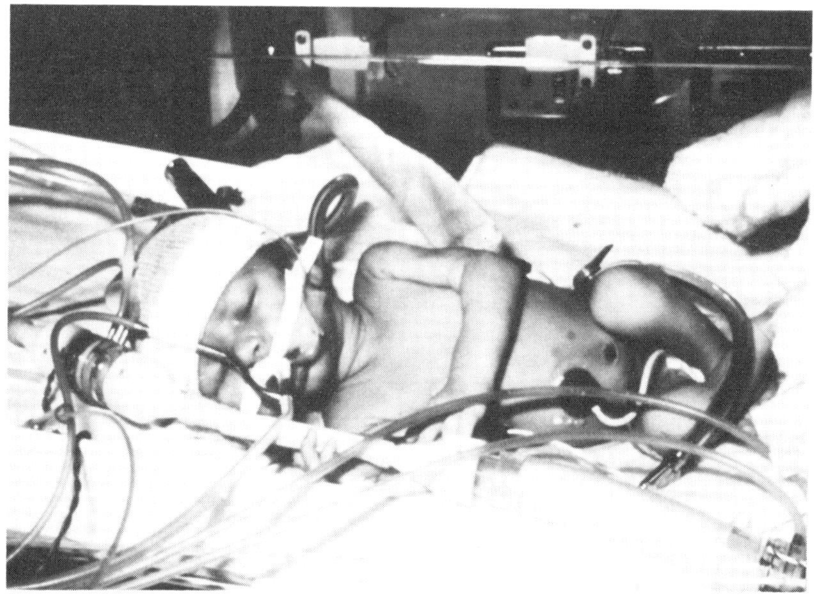

A new-born baby with myasthenia gravis in an intensive care unit.

by producing narrowing of the main arteries supplying the kidneys in rabbits. The consequences have been far reaching: the treatment of high blood pressure, the alleviation of the formerly lethal condition of malignant hypertension, the control of mild to moderate hypertension and the consequent prevention of stroke and the improvement in the prognosis of heart disease have thus depended on animal experimentation. ACE inhibitors are also proving effective in the control of diabetic kidney disease and other forms of chronic nephritis. These pharmacological advances have altered the nature and shape of medical practice.

A major challenge to medicine is the prevention of disease and the greatest triumph of modern medicine is the eradication of smallpox. This achievement had its roots in the empirical observations that milk maids infected with cowpox were resistant to smallpox. The vaccine, however, had to be generated in animals. The movement in modern medicine is away from the empirical and towards the rational. The pharmaceutical industry has shifted emphasis from the screening of large numbers of compounds to the rational design of compounds based on the knowledge of the structure and functions of receptors and their ligands. All of this has been given tremendous pace by the developments in molecular and cell biology.

There is now an international programme to map and sequence the human genome and, in parallel, a similar initiative to map and sequence an animal genome, the mouse, and other species too. A major justification for a programme of this sort is that it will in due course give rise to a new pharmacopoeia. Genes and their products will be identified whose physiology will give new insight into disease and its treatment. Molecular and cell biology has as one of its attractions that it is a reductionist approach to the analysis of biological and medical problems. It is providing for biologists scientific tools comparable with those enjoyed in the past by the physical

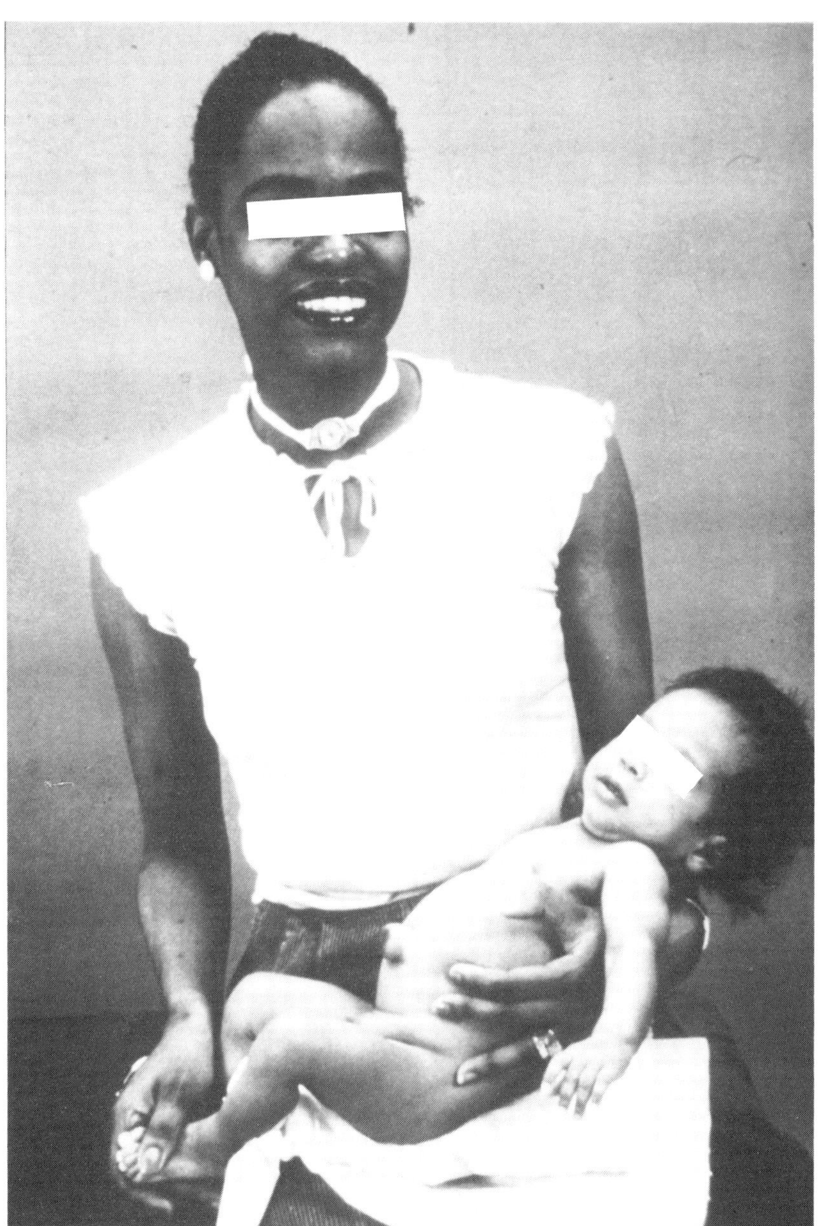

The same baby in Figure I after treatment.

scientists. However, it must be emphasized that the full understanding of biological systems cannot be achieved by the reductionist approach. The advances provided by molecular and cell biology will, of necessity, need to be studied in the whole animal. Nowhere will this principle be more evident than in the study of the nervous system. Diseases of the nervous system cause an enormous burden of ill health in our society, ranging as they do from disorders of behaviour, of addiction to alcohol or smoking, through to the major psychotic diseases such as schizophrenia and manic depressive illness.

In the literature of this Society there are comprehensive accounts of the triumphs of experimentation in animals and it would be redundant for me to spend much time elaborating on these. Nevertheless, as a practising doctor it seems appropriate for me to emphasize the difference between medicine as it is now and the problems that afflicted society in relatively recent times.

During the early part of this century there were a number of major medical advances. Two stand out: the discovery and therapeutic uses of vitamins and the treatment of insulin-dependent diabetes. It is the latter, the treatment of diabetes by insulin, that I would like to consider more fully, because I believe there are lessons to be learnt from this discovery which pertain today.

This story is familiar. The initial observation was made by Minkowski and von Mering (1890) who showed that removal of the pancreas in the experimental animal led to diabetes mellitus. This was followed shortly by the histopathological observation by Opie that in diabetes there were changes in the cells of the islets of Langerhans, which we now know are the cells that produce insulin. This work led to a series of experiments in which attempts to extract the postulated hypoglycaemic factor from the pancreas and use it for the control of experimental diabetes, and thence to try and treat human diabetes, failed. This led to an interesting period in medical history when the American physician Frederick Allen propounded the theory that the problem in diabetes was that patients could not cope with diet and therefore had to be starved.

I thought it might make interesting reading to quote from the excellent book by Michael Bliss (1983) on the discovery of insulin:

> 'In 1913 Frederick Allen, the proponent of the starvation theory pronounced: "all authorities are agreed upon the failure of pancreatic opotherapy in diabetes…injections of pancreatic preparations have proved useless and harmful. The failure began with Minkowski."'

Then Bliss goes on to describe Allen's method of treatment. Another quotation highlights the suffering of insulin-dependent diabetics in that era:

> 'The ironies, the Hobson's choices, the catch-22s of the treatment were staggering. An adult diabetic, weak, emaciated, wasted perhaps to 90 lbs would be brought into hospital and ordered to fast. If the patient or the patient's family complained that he or she was too weak to fast, Dr Allen replied that fasting would help the patient build up strength. If the patient complained about being hungry, Allen said that fasting would help ease the hunger. Suppose the method didn't seem to work and the symptoms seemed to get worse. The answer, Allen insisted, was more vigorous under-nourishment: longer fasting, a maintenance diet even lower in calories. To top it off, Allen and others were also urging diabetics to take as much physical exercise as possible, claiming it would help them burn more food and increase in strength.'

There is one other anecdote which makes that era of medicine come alive. On the question of dietary supervision:

> 'Even inside the hospital the staff had to be constantly on the alert to stop pilfering of "forbidden food". The most extreme example was case four, a 12-year-old boy whose diabetes had already caused blindness when he was admitted. No matter how carefully he was treated, his urine tests on some days would show sugar. It could not be accounted for from his diet. The staff could not understand what was happening. It had seemed that a blind boy isolated in a hospital room was so weak that he could scarcely leave his bed would not be able to obtain food surreptiously when only trustworthy persons were admitted. It turned out that this supposed helplessness was the very thing that gave him opportunities which other persons lacked.... Among unusual things eaten were tooth-paste and bird-seed, the latter being obtained from the cage of a canary which he had asked for.... These facts were obtained by confession after long and plausible denials. The experience illustrates what great care is necessary if records of diabetic patients are to be vouched for as correct.
>
> The gods had their revenge. Thinking that glycosuria was caused by too high a normal diet, the staff cut the boy's normal food supply further and further. It was too late when they realized their mistake. He weighed less than 40 lbs when he died from starvation.'

That is a historian's account of diabetes before Banting, Macleod, Collip and Best carried out their research in dogs which led to the discovery of insulin. Today diabetes remains common and is becoming commoner; the prevalence is doubling every 20 or 30 years. The disease is a major public health burden. Approximately one in 600 school children develop diabetes. One in six of all cases of acquired blindness is due to diabetes. Approximately half the cases of premature myocardial infarction result from the disease. It is probably also the major cause of end-stage renal failure, requiring dialysis and transplantation, in the western world. The process which leads to the disease is an autoimmune destruction of the islets of Langerhans in the pancreas, but the precise cause of that process is unknown. Thus, the success of Banting and his colleagues was only partial. Many complications continue to occur and the questions now to be addressed are whether some other form of treatment such as islet grafting, or the provision of other substances produced by the islet cells in addition to insulin, will prevent or reverse these.

Research in patients with diabetes has contributed to our understanding of the disease. An example is the study of identical twins conducted by David Pyke and his colleagues in which they demonstrated that approximately 40% of identical twins develop the disease. This powerful human genetic experiment identifies the importance of both genetic and environmental factors in this disease. But the problem is that the study of identical twins is a long and slow process and it takes many years before answers are obtained. Here, the power of animal research is evident since there are inbred strains of mice that develop an autoimmune destruction of the islets of

Langerhans which results in diabetes mellitus. Inbred strains of mice have been an enormously powerful tool in biological research because they are genetically more or less identical.

It is of note that the proportion of these animals that develop the disease varies according to how and where these animals are kept. The same strains of mice in the U.S.A., Canada and Japan, may or may not develop the disease. The study of these animals is clearly a powerful way of identifying environmental factors that may be important in causing the disease in man. It also illustrates what must continue to be an ever present approach in medical research—the identification of problems in patients, the recognition that certain aspects cannot be investigated in man and the construction of an appropriate animal model of disease.

One of the main challenges concerned with animal research is not whether animal research is or is not carried out, but whether such research is appropriate. This to my mind is one of the central difficulties for scientists when they defend animal research. There is an enormous need in pharmacological research, for example, for a better animal model of rheumatoid arthritis. The question is, how is such a model to be developed. (See Chapter 3 by Rees for an elaboration on this important issue.) The advances in modern molecular and cell biology—through an understanding of the genetic basis of disease, the identification of comparable genes in experimental animals and the opportunities provided by transgenic animals—will open the way for better models. However, genetic predisposition to most common diseases, as in the case of diabetes, depends on several genes. In diabetes there are probably around six genes involved in creating the predisposition to the disease upon which environmental factors then act. Thus, there is a substantial challenge and opportunity for experimentalists using the new biology to produce models of disease that approximate more closely to the state of affairs that occurs in humans. This can only be brought about by a continuous programme of human and animal investigation.

Transgenic models of disease are particularly appropriate for studying the development of single gene diseases, diseases where a defect in a single gene is responsible for the condition. These disorders are the subject of much current interest since their treatment by 'gene therapy' seems potentially within the grasp of modern medical research. There are somewhere around 4000 such diseases that afflict mankind. Some of them are relatively common, such as cystic fibrosis and genetic disorders of the haemoglobin molecule of which thalassemia is an example, but many are rare. In general, disorders that are rare are extremely difficult to study. Sufficient numbers of subjects often cannot be accumulated to allow statistically valid conclusions to be drawn from clinical experimentation. Appropriate experimental models are therefore vital for therapeutic advances in this area.

I would like to consider the areas of immunology and transplantation. A signal development, by Roy Calne, was the introduction into dogs of azathioprine, one of the main drugs that made organ transplantation possible in man. The scientific basis of transplantation is immunology. Medawar's pioneering experiments in transplantation immunology involved skin grafting between calves, and showed that immunological tolerance was possible. Skins from dizygotic twins of unlike sex, and thus by definition genetically different, were each completely tolerant of grafts of the other's skin. Achieving immunological tolerance by pharmacological and immunological manipulation remains a central goal of modern immunological research and depends critically on animal experimentation. Because of advances that have already been made organ grafting represents one of the major therapeutic achievements of the latter part of this century. Transplants of the kidney, liver, heart and bone marrow have become routine. More than 100 000 kidney grafts have been performed and some of the grafts have survived more than several decades. Many clinicians believe that supply of donor organs will remain the major limiting factor in transplantation and much current research is devoted to understanding the mechanisms underlying rejection of organs grafted between different species.

Although organ transplantation provided a major stimulus to research in immunology it is now clear that much human disease results from disordered function of the immune system. Two areas are of great importance in clinical practice: allergy and autoimmunity. Autoimmunity underlies the development of common and important disorders such as insulin-dependent diabetes mellitus, diseases of the endocrine glands such as the thyroid, diseases of joints (rheumatoid arthritis) and diseases of the nervous system, including multiple sclerosis. The basis of autoimmunity has been studied extensively in experimental animals and there is now a substantial body of knowledge which should soon be applied in clinical practice. The treatments for graft rejection (which were derived from experimental grafting in animals) have been applied to the treatments of these various 'autoimmune diseases' with varying degrees of success. A problem is the overall suppression of the immune response leaving the patient susceptible to infection. It is expected that research into experimental autoimmunity will provide a means of specifically suppressing an unwanted (auto)immune response leaving the rest of the immune system unaffected.

One such approach involves the use of monoclonal antibodies 'engineered' by gene splicing; this enables the specific binding site to be grafted on to the 'backbone' of a human antibody, a technique pioneered in Cambridge by Dr Greg Winter and his colleagues in the Laboratory of Molecular Biology, and adapted for immunotherapy by Professor Herman Waldmann. It is of relevance that the science underlying the development of this approach came from many roots—from chemistry (the structure of antibodies), from studies of the genetics of antibody production in experimental animals, and from molecular biology, specifically from site-directed mutagenesis, a powerful technology originally developed to elucidate the structure–function relationships of macromolecules.

No review of the significance of animal experimentation in the modern context would be complete without mention of the human immunodeficiency virus (HIV). The virus responsible for acquired immunodeficiency syndrome (AIDS), according to the most recent WHO estimates, is being carried by somewhere around 20 million people in the world, and that number is rising rapidly. In one sense the study of HIV represents a spectacular triumph for modern molecular and cell biology.

The first case was reported in 1981. Within about 4 years this virus had been isolated, the complete gene sequence was known and certain partially effective drugs were available. Yet we are still faced with a major problem. One of the reasons for this is that there is no good animal model for AIDS. There are viruses which are similar to HIV which occur in primates (simian immune deficiency virus) and in cats (feline immune deficiency virus), but the only other species that is infected with HIV is the chimpanzee, and in this species the infection does not develop into the full-blown AIDS syndrome.

It is a truism that the control of infectious disease has largely rested upon two foundations: the ability to culture an organism and the availability of an experimental model of the disease. It is significant that the bacterium that causes leprosy was discovered by Hansen (known as Hansen's bacillus) in 1873, but it has not yet been cultured, and only comparatively recently has an animal model become available. This is the nine-banded armadillo which unfortunately cannot be bred in captivity. Thus we have made little progress in providing an effective cure for leprosy. But the significance of this disease is now dwarfed by the problem of AIDS. Because of the lack of an appropriate natural model of AIDS, inbred strains of mice with genetically impaired immune systems grafted with human lymphoid tissue are being studied to provide information which cannot, at present, be obtained by any other means.

Another challenge for medical science today is the malfunction of the nervous system. It is impossible to envisage any way of dealing with the problems of disease of the nervous system without animal experimentation (see Chapter 8 by Hubel). However, it is relevant to emphasize how substantial the advances have been in the control of diseases of the nervous system. Progress includes new treatments such as lithium for manic depressive illness and various neuroleptic agents for schizophrenia, the successful application of anti-depressants and analgesics, the developments in anaesthesia which have made modern surgery possible and the amelioration of Parkinson's disease. These cases represent just a start, for the scale of suffering associated with diseases of the nervous system is huge and it is clear that we need new approaches and further experimentation.

To conclude, I should emphasize that I have made no attempt to be comprehensive. For example I have not considered the substantial improvements in the treatment of certain forms of cancer, especially leukaemia, lymphoma and related disorders which have taken place in the past two decades. Patients today are already experiencing substantial benefits from the explosive growth of knowledge in the biomedical sciences. Much more should be possible, however, as the advances in biology are applied to human disease. This cannot occur without animal experimentation.

The quotations on pages 8 and 9 are reproduced from 'The Discovery of Insulin' by Prof. Michael Bliss with the kind permission of the author.

References

Bliss, M. (1983) The Discovery of Insulin. Paul Harris Publications, Edinburgh
von Mering, J. and Minkowski, O. (1890) Diabetes mellitus nach pankreasextirpation. Naunyn Schmiedebergs Arch. Exp. Pathol. Pharmakol. 26, 317–387

3

Animal experimentation in fundamental research

David Rees
Secretary, Medical Research Council

Introduction

Research with experimental animals does not provide the only major route to advances in biological understanding and delivery of medical benefits, but it does provide an approach that is very important indeed, and one that is likely to remain so for the foreseeable future. In the previous chapter Professor Peters has given us an excellent account of the development of animal experimentation in medical research. It thus seemed appropriate that I should follow on to describe the future direction. I have therefore tried to take stock of the MRC's present and future dependence on animal experimentation for medical research. I have mostly left aside the large and important area of neurosciences and of cancer because these will be covered by Professor Hubel and Dr Bodmer (Chapters 8 and 9), but the general conclusions may well be similar for those areas.

My conclusions are paradoxical. On the one hand, techniques of molecular and cell biology have so greatly improved in power and scope that, in many situations, the use of animals can be sharply reduced or even eliminated altogether. On the other hand, and at the same time, those same developments in molecular and cell biology are opening up important horizons for medical research which will involve new types of animal experimentation; other types of animal use are increasing as a consequence.

This chapter therefore looks first at one side of the coin and then at the other, giving different examples. Firstly, some developments are outlined that have caused animal usage to dwindle. Secondly, new types of animal experimentation are described which offer novel types of progress towards medical benefit and research understanding.

Beyond animal experimentation

One area in which animal usage might become progressively superseded is the preparation of antibody-like molecules. Antibodies have very many uses in medical diagnosis and treatment, based on their ability to bind and lock onto other molecules such as those of disease agents. Over the past 10 to 15 years, they have been made with ever greater precision, and their uses have exploded so that whole new industries have come into being to make and distribute them. However, they have always had to be made by immunization of animals. An important new approach developed by Greg Winter of the MRC Laboratory of Molecular Biology in Cambridge (McCafferty et al., 1990; Winter and Milstein, 1991) now eliminates the need for animals, except in the initial preparation of a sample of DNA which, in principle, need be prepared only once.

The method uses a bacterial virus or a bacteriophage which is a very small, long, thin structure as shown in Figure 1 (Gray et al., 1981), in a

Figure 1

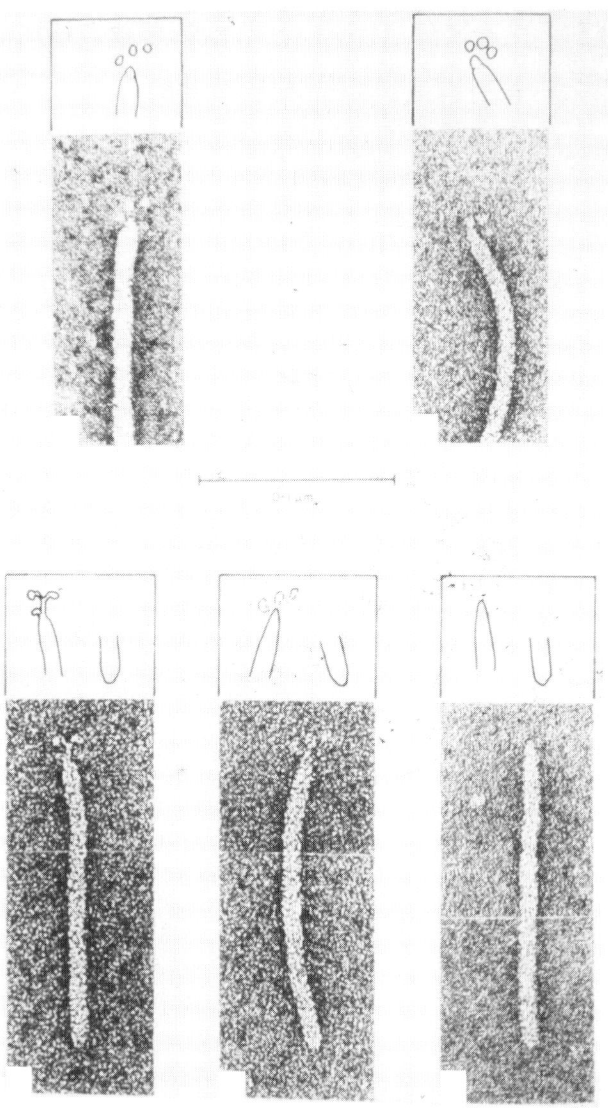

Electron microscope image of the fd virus, negatively stained as an adsorption complex. The different frames show various aspects of the fine structure. From Gray *et al.* (1981) with permission.

preparation shadowed for the electron microscope. It has a core of DNA wrapped in a protein coat; at the top of the rod, a specialized structure is visible which functions to attach to the bacterial host to initiate infection.

Figure 2 (upper) shows a schematic representation of the DNA core surrounded by its protein coat, having a different protein (colour coded green) associated with the attachment structure at one end. When the phage replicates in the host cell, not only is the DNA replicated but its genetic code is translated to make coat protein for the assembly of complete new phage

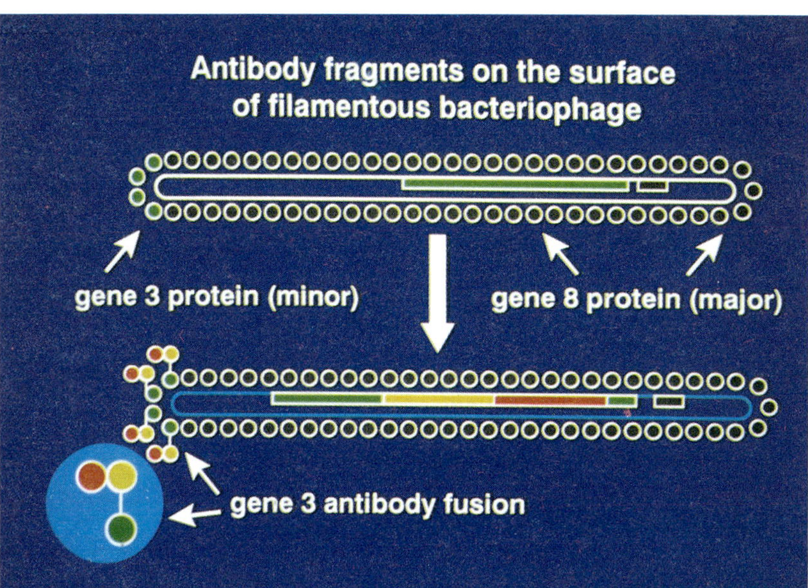

Figure 2

Schematic representation of the organization of the same bacterio-phage as in Figure 1 (upper), and of the organization of the bacteriophage modified to include antibody genes expressed on the surface (lower). *See text for details.*

particles. The Winter technique is to split the gene for the attachment protein (colour coded green) to insert new DNA coding for antibody chains (colour coded yellow and red) so that replication of the new phage now makes particles with antibody chains (coloured yellow and red) joined onto the attachment apparatus. By using a mixture of DNA coding for all possible antibodies, a mixture of phage particles is obtained, each one carrying different antibody chains. The one required can be picked out in the same way as the natural phage picks out its host bacterium. To select antibody against a cancer molecule, for example, the mixture of phage particles is simply presented with the cancer molecule and only the phage particle with the right antibody is bound. One antibody can be selected from a million in this way. The purified bacteriophage expressing the desired specificity can then be grown in unlimited quantities by infecting bacteria for mass fermentation. This method is not only simpler, faster and cheaper than those involving animal use but, in principle, can also give access to a wide range of more tightly defined selectivities. However, it would have been quite impossible to conceive of this approach and develop it were it not for the earlier work which did employ animal experimentation.

Animal usage is also dwindling because of advances in the physical sciences. When faced with the problem of identifying a new chemical molecule, physical scientists could at one time only achieve this by chopping it into pieces for identification so that a picture of the original structure could be built up by fitting these fragments together jigsaw-fashion. The original material was destroyed in the process, which was highly regrettable if, as was usually the case, months or years were required for its preparation. Much faster and easier nowadays are the so called non-destructive methods which use the characteristic pattern of radiation energy frequencies absorbed or emitted by molecules. A wide range of complementary techniques are available, using visible light, ultraviolet or infra-red, all of them either polarized or non-polarized, or radiofrequency radiation. The structure can be deduced by well defined and rigorous rules and without affecting the substance at all. Many of these spectroscopic methods are now being developed into imaging techniques to be applied to living animals and humans; they can be non-destructive to living creatures just as spectroscopy is to molecules. Some imaging instruments are well known and indeed in routine clinical use, particularly computerized axial tomography (CAT) and ultrasound scanners but these are based simply on generalized differences in transparency to particular types of radiation, and not to the sharp and selective absorption at specific wave-lengths that can now be put to use.

For example, certain magnetic resonance imaging techniques developed by Ian Young of the MRC Magnetic Imaging Group at the Royal Post Graduate Medical School, Hammersmith, are based on the principle that protons in water absorb radiofrequency radiation at a slightly different frequency than the protons in fat; this in turn arises because in water the protons are bound to oxygen and in fat they are bound to carbon, giving rise to different degrees of shielding by associated electrons. The signals also have different intensities, depending on the individual concentrations of the two types of proton. Thus, it is possible (Bydder and Young, 1985) to distinguish a tumour from the adjacent bone marrow above and below it in the human leg (shown in Figure 3) because the tumour contains more water than the bone marrow and the bone marrow contains more fat than the tumour. This has clear potential value in following the disease and assessing treatments. The method can also be extended to other problems. A few years ago, questions about key metabolites from organs such as brain, muscle, liver and heart could only be answered by killing the animal and extracting the metabolites from the organ concerned. The technique of chemical-shift imaging by nuclear magnetic resonance now allows the amount of the metabolite and its spatial distribution to be measured in a living organ. We can expect to see gradually developing uses of this non-destructive approach, with obvious implications for animal use.

Another powerful imaging technique is positron emission tomography (PET) scanning, which is based on the principle that certain radioisotopes that can be safely administered to humans distintegrate in a way that allows the radioactive emissions to be measured by purpose-built three-dimensional counters for constructing, by means of much computing, a three-dimensional map of the distribution of the radioisotope around the sample. The example shown (Figure 4) is of living human brains in work by Richard Frackowiak and colleagues at the MRC Cyclotron Unit at Hammersmith (Brooks and Frackowiac, 1989). The radioisotope has been attached to a molecule that is used by the brain to make the important chemical signalling molecule, or neurotransmitter, dopamine. The study is of Parkinson's disease in which it is known that brain cells responsible for making dopamine degenerate, with the result that the chemical-signalling mechanism that controls movement begins to break down. When the radioactive precursor of dopamine is administered, it is taken up by those cells that make dopamine, where it registers in the image as the yellow features

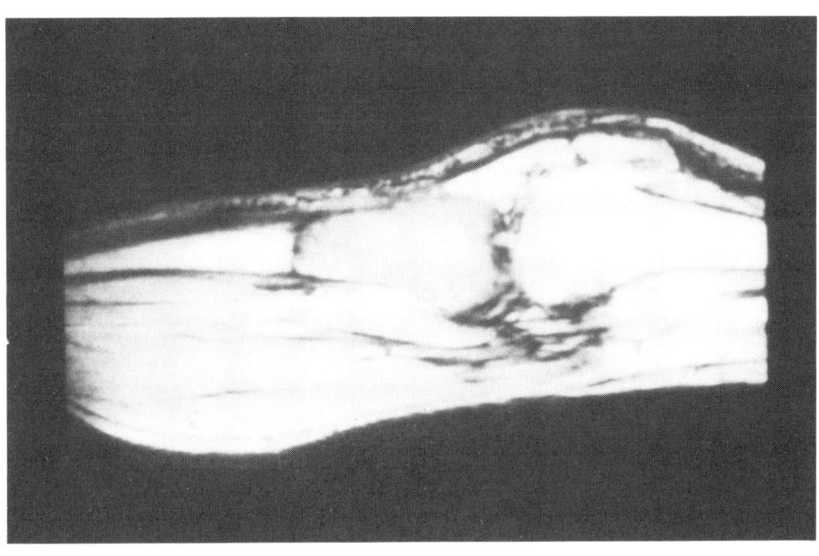

Proton n.m.r. image of human leg (sagittal plane selected). *Chemical-shift differences between water and lipid protons are exploited to show a bone marrow tumour and the normal marrow above and below it. The signals have been processed in such a way as to create a sharp boundary where tumour tissue and normal marrow mix. From Bydder and Young (1985).*

with red centres (Figure 4). The patient with Parkinson's disease (lower pair of images) is clearly less well endowed with functional dopamine-manufacturing cells than the normal subject (upper pair of images). As with magnetic resonance imaging, therefore, we again have an imaging method to follow the course of a disease and evaluate the efficacy of treatments (this time for Parkinson's in the brain rather than cancer in the bone marrow of the leg). The implications for animal use are again obvious. None of these imaging methods could have been developed, however, without access to results from earlier research involving animal experimentation.

Continuing the novel needs for animal experimentation

As mentioned in the introduction, despite the advances which supersede animal experimentation, the need for this type of research will not be totally eliminated in the foreseeable future. A criticism often repeated by opponents of animal experimentation is that:

> 'Animal and human diseases are rarely, if ever, identical. Using an animal model that is not identical to a human disease is a logically flawed process' (Advocates for Animals, 1991).

I shall illustrate why this argument cannot be sustained by means of an example from cancer causation—namely, a recent scare concerning the use of clingfilm for food wrapping. Certain chemicals leaching from this food packaging material were found, in high doses, to produce liver tumours in rats and mice. When the process of cancer development was examined by

Figure 4

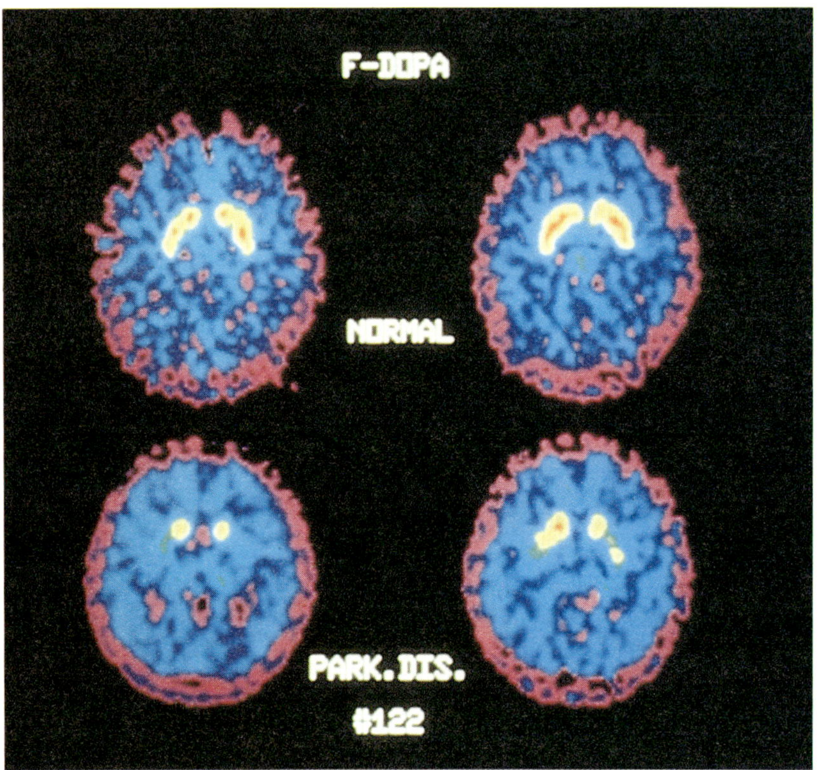

Image of human brain by PET scanning, after the administration of dopa (the biological precursor and analogue of the neurotransmitter dopamine) labelled with the radioactive fluorine isotope, ^{18}F. *The yellow features with red centres show clearly defined substantia nigra; note much stronger labelling in the normal brain (two upper images) than in the brain with early onset of Parkinson's disease (lower pair of images). See text for details.*

electron microscopy, it was seen that certain changes occurred in the liver before the cancer (Figure 5; Keith et al., 1988). The affected liver clearly shows a proliferation of cellular components known as peroxisomes, little factories for breaking down fat and other components taken up from foodstuffs. As a by-product of this activity they produce hydrogen peroxide which is tolerable in moderation, but which in excess can cause damage to the DNA and lead to cancer by a mechanism that has been established previously.

Although the development of cancer could only be studied in animals the initial peroxisome proliferation could be studied in rat or mouse liver cells in tissue culture as well as in the livers of intact animals. In contrast to rats and mice, guinea pigs neither developed cancer with high levels of the additive, nor showed peroxisome proliferation. Not only does guinea pig liver fail to develop cancer, therefore, but its liver cells—both in the intact animal and in cell culture—even fail to get to the first base of peroxisome

Figure 5

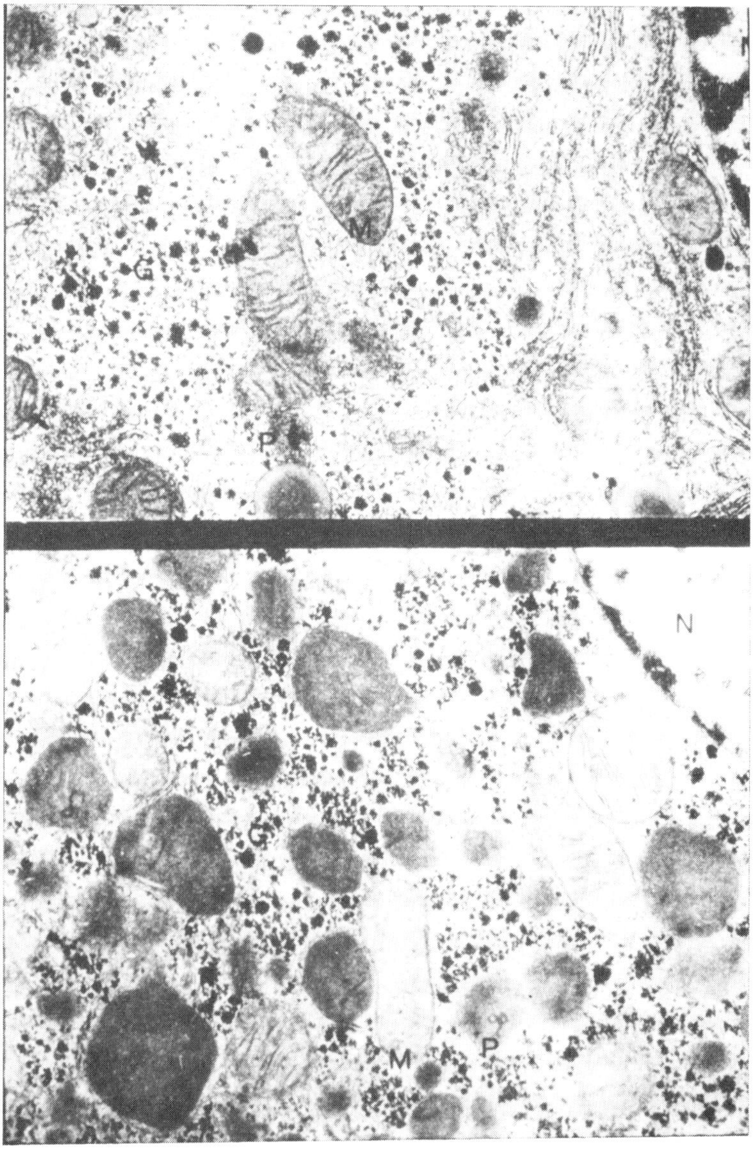

Electron microscopy image of mouse liver in thin section, before (top) and after (bottom) feeding the animal with (di-2-ethylhexyl)adipate, a plasticizer used to impart the cling and stretch to clingfilm. *Since this additive is not covalently bound in the film, it is able to migrate into fatty foods. Note the greater abundance of balloon-like structures known as peroxisomes (P) in the lower panel, clearly distinguished from mitochondria (M) which are of similar dimensions but have a striped internal structure and are often elongated in shape. G, glycogen; N, nucleus.*

proliferation. Human liver cells derived from accident victims were also found not to display peroxisome proliferation, showing that the human cells also fail to achieve first base with the particular additives in question, and therefore that the human liver is unlikely to develop cancer from this cause.

As the critics say, we cannot take it for granted that any animal model will faithfully model the specific aspects of human biology we wish to study. Rats and mice do not do so in this example. However, the studies in cell culture indicated that the guinea-pig *is* a useful model for this particular problem. Although animal experiments were inconclusive at first, cell culture showed which animal model was relevant and which was not. Any responsible research worker will always dig deeper into the mechanisms to check those elements of the comparison between his model and the human which correspond and distinguish them from those that do not.

The importance of using the right animal model also explains the need to use the particular animals most closely related to man in some circumstances, an example being the use of primates for testing the efficacy of polio vaccines or of contraceptive treatment. Another example is from recent MRC work directed at the development of a vaccine against human immunodeficiency virus (HIV), the virus responsible for acquired immunodeficiency syndrome (AIDS). As Keith Peters mentioned in the previous chapter, little progress is being made in using HIV itself in animal models, but some species of monkeys in the wild for example the African Green Monkey (Figure 6) are known to carry the simian immunodeficiency virus (SIV). A number of strains of this monkey virus have been isolated and studied in detail and shown to bear a close resemblance to HIV in terms of physical and genetic structure, immunological characteristics and pathological effects both on isolated cells and in whole animals. Through the investigation of this infection in monkeys, we can establish important leads for research into the human disease. In particular, MRC workers at the National Institute for Biological Standards and Control and at the Centre for Applied Microbiology and Research at Porton (Stott et al., 1990) have shown that a vaccine can be made by chemical inactivation of infected cells, which protects the animal against subsequent challenge with live SIV. The protection even extended to challenge with a quite distinct SIV strain, providing powerful encouragement that the greater variability of these viruses need not make the traditional vaccine approach unworkable. It will now be possible to identify the particular molecular components of the virus essential for protection, to determine the best route and schedule of vaccination and discover the other components that might be added to improve the vaccine (the so-called adjuvant problem), and to study possible adverse effects or dangers that might be relevant to the immunization of humans.

Recent developments have led to new types of non-primate models which closely parallel human conditions and therefore allow further economy in the use of primate species. One example—also highly relevant to AIDS research—is the so-called SCID-hu mouse, derived from a freak genetic strain, called severe combined immunodeficient (SCID) mice, which arose when mice were repeatedly bred in an absolutely germ-free environment. In the absence of any infectious agent in their environment, a new strain originated which had lost its immune system. Immunological rejection thus being impossible, these mice can take tissue transplants from other species including human. As an example, they can be populated with human white blood cells or lymphocytes (see Figure 7). Since lymphocytes are the

Figure 6

African Green Monkey (*Cercopithesus ethiops*) which is indigenous to many parts of Africa and harbours SIV which can cause a disease closely resembling AIDS in certain species of monkey.

targets for HIV, these human cells can be infected in these mice and the progress of the disease studied. It turns out that the consequences of HIV infection of human cells in the mouse parallels that of the human cells in humans, at least in some respects (Figure 8; Mosier et al., 1991). T-cell counts decline (as shown by the filled bars) as they do in humans, and they do so at a rate that varies from individual to individual. Immunoglobin levels also decline, in a way that correlates with the decline in the T-cell drop (again as in humans), perhaps because the depleted T-cells include those that regulate immunoglobin synthesis. Despite a very hostile report about these mice and about their commercialization (Tyler, 1991), it is believed by the MRC that they are likely to prove of real practical value in the fight against AIDS, perhaps for testing anti-viral agents, both for efficacy and for toxicity evaluation. As with all animal models, of course, it will be essential to watch and test carefully any limits to validity as a true model of the human disease.

The SCID mouse is useful because it can act as a host for the parts of the human physiology we wish to study. A further stage in sophistication will be the incorporation into other species of specific human genes coding for individual proteins such as those that cause disease. The whole sequence of events that follows gene expression can then be studied right through to disease symptoms, hence elucidating the disease mechanisms and enabling the discovery of possibilities for new therapeutic intervention. Even beyond this

Figure 7

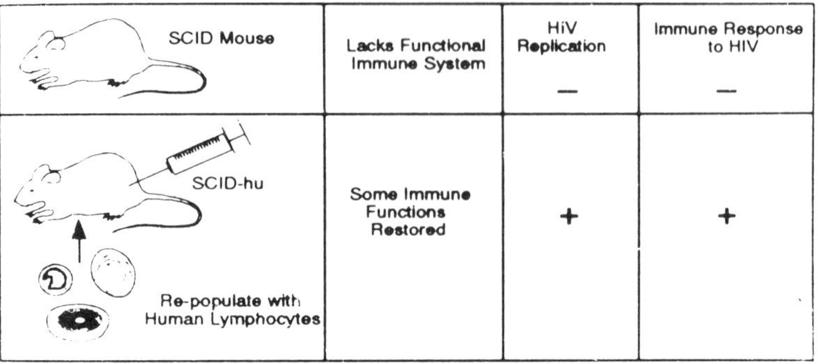

Diagrammatic representation of the principles of use of the SCID-hu mouse for AIDS research. *Mice with congenital severe combined immunodeficiency are repopulated with human lymphocytes, which are then susceptible to infection by HIV.*

Figure 8

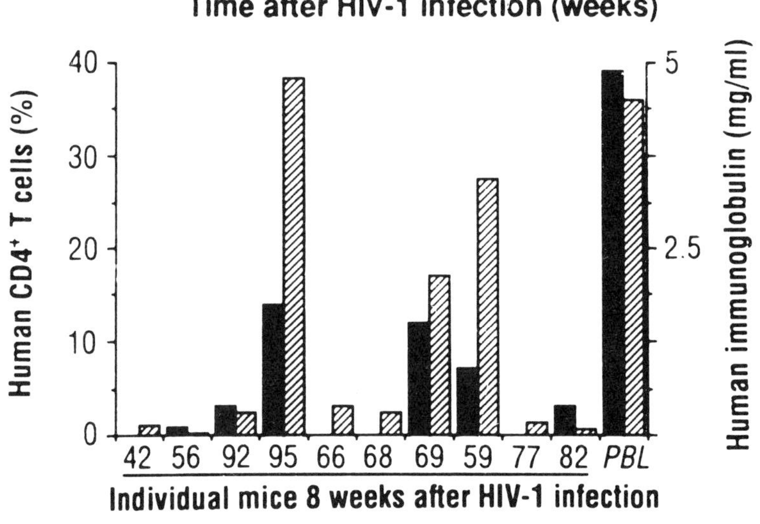

Alterations in CD4+ T-cell percentages (filled columns) in the spleens and human immunoglobulin concentrations (striped columns) of individual SCID-hu mice infected 8 weeks earlier with HIV. *PBL, control values. Adapted from Mosier et al. (1991) with permission. Copyright 1991 AAAS.*

stage, the model offers a system for testing and evaluating possible treatments. One example is from recent studies by Grosveld and his colleagues at the MRC's National Institute for Medical Research at Mill Hill, who have constructed a mouse model of human sickle cell anaemia (Greaves et al., 1990). This inherited disease, which is a significant public health problem wherever there is a substantial black population, is carried by a single point mutation in the β-chain of haemoglobin, which results in a single amino acid replacement—valine for glutamic acid. This small change in structure causes a very large change in the properties of haemoglobin. In its deoxygenated form, the altered haemoglobin polymerizes within the red blood cells to form fibres which distort cell shape. As a result red cells do not survive their normal life span and anaemia results. Their shape also impedes movement through smaller blood vessels; hence, blockage and tissue damage result from the restricted blood flow. To make transgenic mice (Figure 9), the foreign DNA is injected into the nucleus of fertilized eggs which are then returned to a foster mother which eventually gives birth. Those progeny that express the transfected gene are then bred to produce so-called homozygotes, which express two copies of the transgene. Not only the altered gene itself must be introduced into the experimental animal, but also other DNA sequences which carry the instructions to ensure that the protein product is expressed specifically in the right place—in this case in blood cells rather than, for example, brain cells or heart cells. This was possible because the team led by

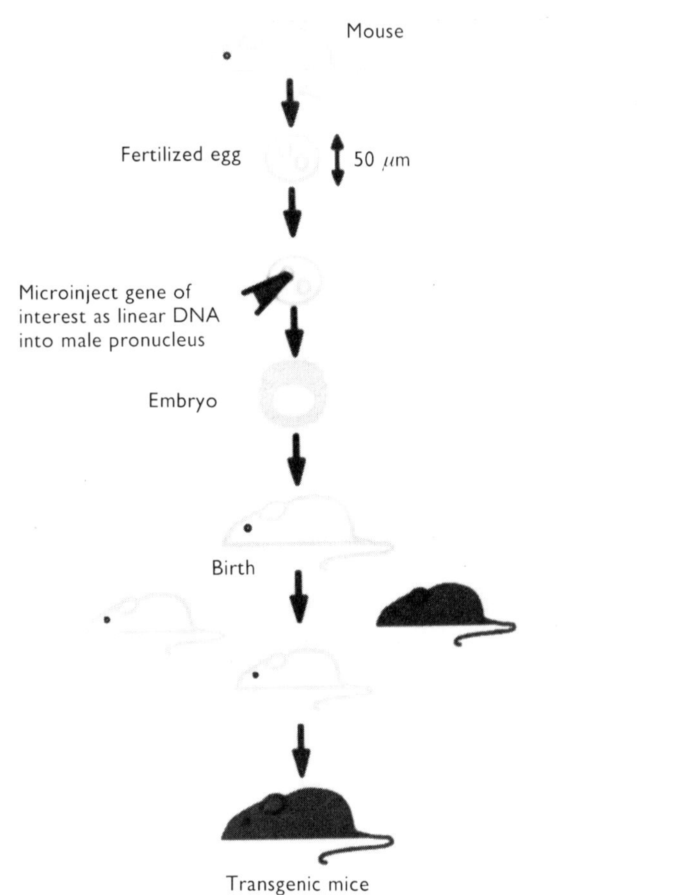

Figure 9

A standard procedure used to make transgenic mice. *In this example, the gene injected into the mouse egg causes a change in coat colour.*

Grosveld had made crucial discoveries about the factors that control such expression of haemoglobin genes (Talbot et al., 1989). Those mice expressing the sickle cell gene had red cells which, under reducing conditions to deoxygenate the haemoglobin, were sickle shaped (like diseased human cells) due to the polymerization of the haemoglobin (Figure 10).

This now opens the way to develop treatments and even cures for sickle cell disease in humans. One approach could involve replacement of the defective gene. Since red cells are constantly regenerated from the bone marrow and bone marrow transplantation is an operation that is already well established, patient's bone marrow with the defective globin gene could be removed. The aim would be to replace the defective gene by a correct gene and return the bone marrow to the patient. It is not known exactly how this might be achieved, but there are various approaches available for exploration, and the transgenic mouse model can be used for these tests and to optimize any selected for use in humans. Another possible approach is less direct. Instead of aiming to replace the defective gene in a hospital operation, a drug might be designed to control aggregation of the defective protein; the feasibility of this also could be explored in the experimental animal and then optimized prior to clinical trials.

There is great scope for developing this type of approach to the study of other genetic diseases that afflict man, and research activity is considerable worldwide. It seems difficult to overestimate the benefits for medicine and biological research which will follow.

Another recent example of transgenic construction of a human disease model concerns the unpleasant disease known as ankylosing spondylitis, a chronic inflammatory disease affecting bones and joints. As the disease progresses, fusion of the joints can occur such as the joining of adjacent vertebrae resulting in severe curvature and the classic 'bamboo spine' (Figure 11).

The great majority of affected individuals have inherited a particular feature of the so-called histocompatibility system, whose function is to recognize foreign molecules for triggering of the immune defence system. A specific gene form or allele, known as HLA-B27, is associated with the disease and it seems that the problems arise from the interactions of the

Figure 10

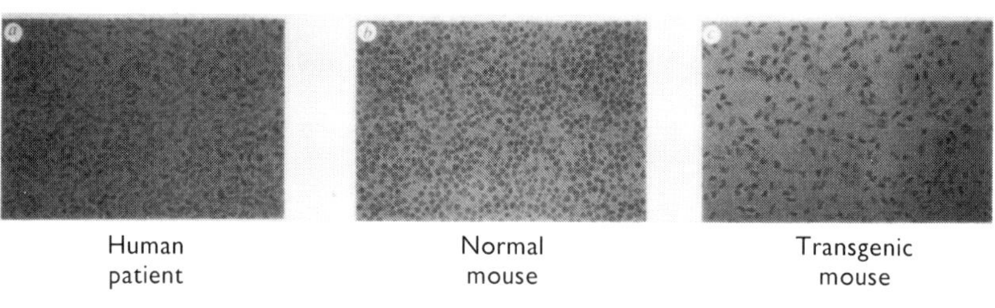

Erythrocytes incubated under reducing conditions (sodium metabisulphite) to show that sickling occurs in a human patient with sickle cell disease (a) and in the transgenic mouse carrying the disease gene (c), in contrast to the normal mouse which shows no such sickling (b).

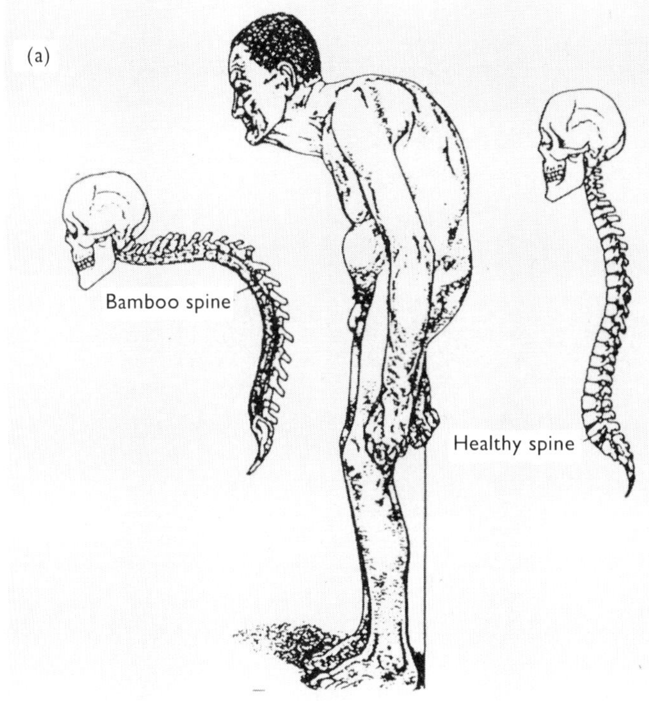

The stooped posture seen in ankylosing spondylitis patients with severe disease. *As the disease progresses, syndesmophytes form between the vertebrae resulting in the classic 'bamboo spine'. From 'The Independent', 18 November 1988.*

product of this gene with certain types of bacterial infection (*Klebsiella*), which causes a severe inflammatory response as a side effect and in turn produce further problems.

Attempts to develop an animal model by introducing this gene into transgenic mice, as described above, have been unsuccessful. A group from the University of Texas (Hammer et al., 1990), however, reasoned that the experiments might have failed in mice because of unknown features of physiology differing crucially from the human, and that better results might be obtained from another species (in short, because mouse is a poor model for human here). They turned to the rat because it was known that rats are more susceptible to several experimentally induced bone diseases that cannot be elicited in mice. Such transgenic rats did indeed spontaneously develop an inflammatory disease very similar to the human disorder. For example, the transgenic rat shows severe curvature of the tail compared with the normal rat (Figure 12), resembling the curvature of the human spine (Figure 11). It can be concluded that the HLA-B27 allele plays a central role in the disease. Detailed cellular and molecular analysis of the rat model will now lead to better understanding of the human disease and again provide a test bed for the development of new drugs.

Figure 12

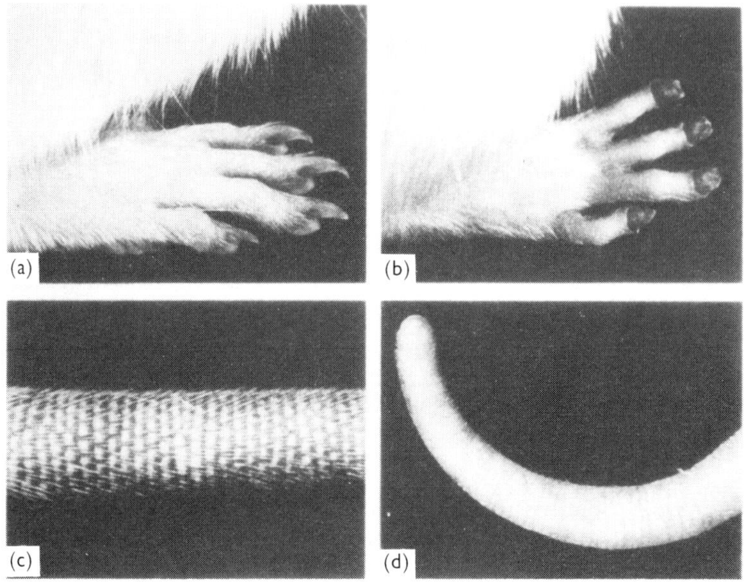

Pathology of the spontaneous inflammatory disease in transgenic rats expressing HLA-B27, as a model for human ankylosing spondylitis. *(a) Normal hindlimb digits and nails. (b) Hindlimb digits and nails of a $3\frac{1}{2}$-month old 21-4 male, showing hyperkeratosis and dystrophy of the nails and alopecia over the digits. (c) Normal tail. (d) Tail of a $3\frac{1}{2}$-month old 21-4 male [same as in (b)], showing oedema, alopecia, flaking and masking of the normal ridged pattern. Adapted from Hammer et al. (1990) with permission. Copyright Cell Press 1990.*

Conclusions

To those hoping that I would be able to point to a future when fundamental medical research would have no further need of animal experimentation, my message might perhaps be disappointing. I end by repeating the conclusion proposed at the beginning. On the one hand, techniques of molecular and cell biology have so greatly improved in power and scope that, in many situations, the use of animals can be sharply reduced or even eliminated altogether. However, this would not have been possible were it not for earlier results obtained from animal experimentation. On the other hand, new types of animal experimentation are opening up important horizons for medical research, with the result that other types of animal use are increasing.

Animal usage in research funded by the MRC has declined substantially over the past few years. Since 1984, our usage of rats and mice has at least halved and that of domestic species (cats, sheep, pigs, goats) has reduced by much more than that. The employment of primates has always been low but, despite new AIDS research, is drastically down on former years. In our establishments, we currently use mainly rats and mice (about 100 000 a year), but also around 1000 rabbits, 1000 chickens, birds and frogs, 50 sheep, pigs and goats, and a similar number of cats and monkeys. The numbers may well dwindle further, but will not vanish. Animal experimentation is changing in its pattern rather than disappearing. As with so many ethical questions that confront the human conscience, the debate on whether to continue with animal experimentation shows no sign of being resolved by scientific or social progress. Many of the types of experiments we talk about now are different from 20 or even 10 years ago, but they simply put the old arguments and the dilemmas in modern clothes. Ethical issues here change their form rather than their essence, just as they do in crime and punishment or love and war. If human and indeed animal life is to be sustained on this planet, let alone improved in quality, then medical and health research must continue. Animal experimentation is an inseparable part of the seamless web of this research process and therefore must also continue. This must be done with all respect and consideration for the animal subjects themselves.

References

Advocates for Animals, 79th Annual Report (1990), p. 15

Brooks, D. J. and Frackowiak, R. S. J. (1989) PET and movement disorders. J. Neurol. Neurosurg. Psychiat, Special Suppl., 68–77.

Bydder, G. M. and Young, I. R. (1985) Clinical use of the partial saturation and saturation recovery sequences in MR imaging. J. Comput. Tomog. 9, 1020–1032

Gray, C. W., Brown, R. S. and Marvin, D. A. (1981) Adsorption complex of filamentous fd virus. J. Mol. Biol. 146, 621–627

Greaves, D. R., Fraser, P., Vidal, M. A., Hedges, M. J., Ropers, D. and Grosveld, F. (1990) A transgenic mouse model of sickle cell disorder. Nature (London) 343, 183–185

Hammer, R. E., Maika, S. D., Richardson, J. A. and Tang, J. P. (1990) Spontaneous inflammatory disease in transgenic rats expressing HLA-B27 and human β-2M: an animal model of HLA-B27-associated human disorders. Cell 63, 1099–1112

Keith, Y., Cornu, M. C., Elcombe, C. R. and Lhuguenot, J. C. (1988) Di-(2-hydroxyethyl)adipate and peroxisome proliferation: identification of proximate peroxisomal proliferator *in vivo* and *in vitro*. The target organ and the toxic process. Arch. Toxicol. Suppl. 12, 274–277

McCafferty, J., Griffiths, A. D., Winter, G. and Chiswell, D. J. (1990) Phage antibodies: filamentous phage displaying antibody variable domains. Nature (London) 348, 552–554

Mosier, D. E., Gulizia, R. J., Baird, S. M., Wilson, D. B. and Spector, S. A. (1991) Human immunodeficiency virus infection of human-PBL-SCID mice. Science 251, 791–794

Stott, E. J., Chan, W. L., Mills, K. H. G., Page, M., Taffs, F., Cranage, M., Greenaway, P. and Kitchin, P. (1990) Preliminary report: protection of cynamolgus macaques against Simian immunodeficiency virus by fixed infected-cell vaccine. Lancet 336, 1538–1541

Talbot, D., Collis, P., Antoniou, M, Vidal, M. and Grosveld, F. (1989) A dominant control region from the human β-globin locus conferring integration site-independent gene expression. Nature (London) 338, 352–355

Tyler, A. (1991) Mice as people. The Independent Magazine, 20 April, pp. 26–30, London

Winter, G. and Milstein, C. (1991) Man-made antibodies. Nature (London) 349, 293–299

4

Discussion

Sir Michael Woodruff (Edinburgh): I would like to make a couple of points relating to issues raised by Professor Peters. The first, which concerns transplantation, clearly illustrates the point made by Dr Rees, namely, that it is important to chose your animal model. The development of transplantation has been greatly dependent on using animals of a kind which seem to arouse more opposition, in some quarters anyway, than the familiar laboratory mouse. I remember Peter Medawar once saying that he saw no application clinically of research in his laboratory and elsewhere on transplantation and immunological tolerance. However, when surgeons began using organ transplants in larger animals, instead of skin grafts in mice and rats, Medawar was found to be wrong, because it proved to be much easier to prevent rejection of an organ allograft than an allograft of skin. Professor Peters did not mention xenografts (i.e. grafts from one species to another) but, at a recent meeting of the British Transplantation Society in Cambridge, there were many papers on the experimental use of xenografts and it was seriously suggested that xenografts of organs might one day have a place in clinical practice. If this were to happen it would pose its own series of ethical problems.

My other point concerns the development of open heart surgery, pioneered by John Gibbon in Philadelphia, which Professor Peters mentioned briefly. It is inconceivable that this could have been developed without animal experiments—animal experiments, moreoever, in species which have tended to arouse particular opposition in some quarters.

Michael Festing (MRC Toxicology Unit): I have been concerned with laboratory animals for about 25 years and, over that time, I have reached the conclusion that scientists are concerned and caring. But the problem is that they are also in some respects ignorant, and animal experimentation is now so sophisticated that some scientists actually lack the training and information to make the best use of it.

Let me give you a specific example. More than 50 years ago R. A. Fisher wrote his book on the design of experiments and in it he put forward two principles: the principle of blocking to increase the precision of an experiment and the principle of factorial arrangements of treatments to increase the range of applicability. In agricultural research about 80% of experiments use the principle of blocking to increase precision. One might ask therefore what has happened in animal research. I have recently carried out a survey of approximately 70 papers in two toxicology journals and only 2% use blocking and only 4% use a factorial design. In other words they do not use a modern experimental design.

To get some idea of the consequences of this, I looked back over my own research where I had used blocking in several experiments over the past few years, and the relative precision was between 100 and 500%. In other words, by simply using modern experimental design techniques one can get substantially more information from the same number of animals. I must conclude, therefore, that poor experimental design is leading to a lot of mis-use of scientific resources in some cases and possibly excessive use of animals. I think that what we need in the future is a much more informed use of animals. We need to make more substantial efforts in the training of scientists in the use of animals, as well as supplying much better sources of information. In this respect I am very sad that the MRC closed down the Laboratory Animal Centre a few years ago, which was the one source in this country that provided information on laboratory animals.

How are people to find out nowadays about the 400 inbred strains of mice, the 200 inbred strains of rats, the 1000 mutants of mice, all the congenic strains and all other such relevant information? People need a central source of information and at the moment they have to rely on commercial breeders. Although they do quite a good job, I think is a great shame that, in this respect, this country has retrogressed in many ways over the past decade.

David Rees (MRC): I know there are a lot of hard feelings about the closure of the Laboratory Animal Centre, just as there are about other forced economies that the MRC has had to make. But I really cannot accept that that would be the only source of advice of the type that you have mentioned, about particular genetic strains for example. You will be aware that we are considering how to sustain the expertise and the genetic material that exists in the genetic division of the Radiobiology Unit, for example, which represents a considerable repository of knowledge. There are others as well. As people get more interested in transgenic animals I think very many scientists are becoming more knowledgeable about these matters.

But the particular problem you mentioned apart from the question of information, if I understood you correctly, seems to be specific to toxicology. This ought to be covered by the procedures which Lord Adrian mentioned in Chapter 1, that is by the mechanisms that determine MRC funding and the issue of project licences. The matter of capability is intended to be covered by the conferment of personal licences. I am alarmed if you are suggesting there is a problem in the MRC Toxicology Unit in this respect and, if there is, maybe I had better come down and talk about it next week!

I am also rather puzzled as to what this says about the standard of toxicological journals, since papers that go through a rigorous refereeing process ought not be be as lacking in modern techniques of statistical analysis if they are accepted for publication.

Paul Walker (Muscular Dystrophy Group): A number of us here are not actually scientists. We are in medical research charities and, as you probably know, the British Union Against Vivisection (BUAV) has issued a fairly hefty attack on these charities recently. Part of the reason that the BUAV have switched their attention to us is because we now spend over 180 million pounds a year, between us all, on research.

I would like to add a little mouse story of my own, if I may, which is a part of that attack. In their document which they circulated last year, the BUAV named 10 charities and they asked the public not to support the work of these particular charities. Now in our case the work that they criticized was work on the MDX mouse, and they stated that, in human Duchenne dystrophy, the protein dystrophin is missing and, although the same thing occurs in the MDX mouse, it cannot be a proper model because in patients with Duchenne dystrophy their muscles degenerate and they die in their late teens or early 20s, whereas in the mouse that doesn't happen at all—they are perfectly healthy because the regeneration process is good. Surely that is exactly the reason why we *should* use that mouse—to find out why (in contrast to humans) in that particular case, in that particular species, the muscle can regenerate. Perhaps from that we shall learn how we may in some way combat the human disease. I think that it is terribly important that those of us who are involved in animal research, either as scientists or as funders, get the facts right so that we can answer these kinds of criticisms and claims.

An organization of some of the charities already exists. Founded by the ICRF and called the Research for Health Charities Group, it is currently putting together literature and so on. However, I think that there is a need for further action, such as the meeting between the MRC and the Association of Medical Research Charities (AMRC), which took place a couple of months ago at Leeds Castle, to discuss how the two could work together. But I think the responsibility goes beyond the MRC and the charities. Anyone in universities or anywhere else who is actually involved with the application or the funding of animal research should make their voice heard properly to the public.

Lord Adrian: Can I say how much I welcomed the formation of the Charities Research Group and I will certainly do my best to promote its aims. But could I ask whether there have been any noticeable effects of the BUAV's campaign on charitable donations? Has it actually been effective in their terms?

Paul Walker: In our case we cannot actually trace it, but the situation is rather similar to that witnessed in the embryo research case. Lord Walton, our Chairman, was one of the most forthright speakers in the debate. We lost a number of people and a number of voluntary branches but, in fact, the effect was really minimal. In the present situation I think we have lost two people. Probably the charities will continue to be attacked. I think that the BUAV have turned to us simply because it is easier to say to the public: 'don't give money to the Muscular Dystrophy Group; don't give money to the Cancer Research Campaign—they do all these terrible things to animals.' If that really worked, then we would suffer very much indeed. I find it very hard to assess the effect as yet. There are other charity directors present; some of them may have further information.

Sir Walter Bodmer (ICRF): Speaking for a charity and also being a scientist I would agree with Paul Walker. I think that the effect on the charities has actually been minimal. Sometimes, unfortunately, there is an effect on the volunteers, however. We have charity shops for example which can suffer really rather unpleasant attacks. But I find, as I am sure many of us involved in the charitable field do, that when you describe what animal experiments you are doing and explain why, the support is enormous and the rejection is really absolutely minimal. So far, fortunately, such attacks are certainly not having the effect that the antivivisectionists would hope.

Lord Adrian: Before calling on the next speaker, I have been asked to announce that anyone leaving the building at lunch time should remove their badge while out of the building. I don't know if that has a sinister aspect!

Peter McNaughton (Department of Physiology, Cambridge): I wish to make a point which I think overlaps in a way with what the previous speakers have said and which has in fact been implicit in what all three speakers have said this morning. But I wish to make the point a little more forcefully.

The Animal Liberationists pretend to occupy the moral high ground. They say: 'don't exploit animals for research which is of little or no benefit'. They also imply, and may state openly as well, that scientists simply engage in research to satisfy their own curiosity, rather than for any useful purpose, and that this type of research is not productive, or that it may have been productive in the past, but is not now so.

I was particularly pleased to hear the emphatic statement in the lecture by Dr Rees that animal research will continue to be useful. But I think that scientists have to emphasize even further that this research *will* continue to be useful. I would like to put it even more cogently by expressing it in moral terms. Are people who oppose medical research involving animals actually prepared to deny their own children and themselves the fruits of that medical research? Because this is the only logically consistent moral position to take, and I think that the battle really has to be joined on those moral terms, and it has to be joined at the interface between medicine and the public where it really counts—in the doctors' surgeries. One would hope that doctors would be encouraged to say: 'I am prescribing you a medicine which is the result of animal research' (this would be true for almost every medicine which is prescribed). 'Are you prepared, if you oppose animal research, to forego this medicine that I'm prescribing you?' I am sure that Professor Hubel will address this later on but I wished to emphasize the point at this stage.

David Rees: Obviously there is widespread agreement with your general point, but I would like to pick up the particular point you made about the prosecution of animal research which has no foreseeable, immediate application as also being important and justified. I very much agree with this and it is an issue which, given more time, I would have emphasized in my talk.

Vincent Marks (Chairman, 'Healthwatch'): My organization was set up to try to inform people about the choices in medicine. One of the issues upon which we are constantly approached by the media is the role of so-called 'complementary' or 'alternative' medicine. This is generally not medicine at all; it is often, in fact, charlatanism under another name. Many of the procedures used are anti-scientific and deny the real progress that has occurred in medicine. They are presented as though they were beneficial to the community. Many of us who are approached for our views on alternative or complementary medicine (and they can't really make up their minds whether they want to be complementary to medicine or an alternative to medicine) see this as an anti-scientific approach to healing. I think that we as members of the scientific community have a great responsibility to point out that the scientific basis of modern medicine is one of the main reasons that most people are living such a good life today compared with 200 years ago.

Lord Adrian: I would be happy to agree with part of that but I think that one must be a little more cautious about complementary medicine. Consider, for example, the insistence on not bleeding, fresh air and good diet that was characteristic of homoeopathy in the 19th Century. This issue is not all black and white.

Vincent Marks: Sir, I did intend to distinguish between complementary medicine, which I think does have some role to play, and alternative medicine which is really not medicine at all.

Colin Blakemore (Oxford): I endorse wholeheartedly Peter McNaughton's excellent suggestion that those who oppose the use of animals should be expected to deny themselves the benefit of products and treatments resulting from animal research. However, at the moment they would find it all too easy to say that they *would* so deny themselves, simply because of the level of public ignorance about the extent to which they depend on animal use in all aspects of their lives, and not just medical research. This is largely the result, on the one hand, of propaganda to which they are exposed from the opponents of animal use and, on the other hand, of the absence of reliable information with which to contradict that propaganda.

I will give you two examples. A number of antivivisectionist organizations, sometimes using glossy, widely distributed leaflets and press releases etc., have recently stated, quite baldly, that not only no medical advance, but not even any scientific breakthrough has *ever* resulted from the use of animals in research. Such statements seem frequently to be accompanied by the claim that the AIDS virus was manufactured by animal researchers in their laboratories and other patent nonsense. Although admirably contradicted by the RDS in its own literature, such outrageous statements have gone almost unopposed in the media. Of course we must all protect freedom of speech, but I find it amazing that there should be the opportunity to make statements like that without opposition. One problem is that those best placed to contradict such propaganda often regard it as so non-sensical that it does not warrant objection. Unfortunately this results in the less knowledgeable public seeing such claims repeatedly frequently and contradicted rarely. If there is one useful result of this meeting today, it might be that our own perception of the nature of that problem, the urgent need to communicate, will be elevated. We have little chance of changing the minds of the most radical opponents of animal experimentation, but we can and must persuade the public on this issue because in the end it is public opinion that really matters. Public opinion influences governments and thus legislation: we then bear the fruits of that.

I will give just one other example of the depth of public ignorance on this point. The difficulty I want to highlight is the public's poor perception of the extent to which their lives, their safety and their well-being depend on the results of animal use. The 'Daily Telegraph' commissioned a poll by Gallup about a year ago. One of the questions asked was the following: 'Do you refuse (not *would* you, if you could, but *do* you) on principle to buy goods which have been tested on animals or that contain animal products?' Thirty eight percent of the British public believed that they did! Indeed 49% of British women said 'yes'; they were convinced that they were living without in any way depending on the animal world. This would mean without meat, wool, leather, drugs, vaccines, cosmetics or any domestic product which has been tested on animals. We must be aware of the level of ignorance we have to confront if we wish to make our case adequately to the public.

Anthony Dayan (St Bartholomew's Hospital): Could I return to the question of toxicology which has just been maligned quite unnecessarily by Michael Festing and which has received very poor support in the past year from other organizations, apart from the RDS which has been very helpful indeed. I think that there is a fundamental misunderstanding as to what we mean by toxicology. It consists of two entirely different activities.

The first is refined scientific experimentation of the sort that we have heard discussed this morning. The scientist knows what he wants to do, he knows how to do it and naturally he will attempt to control everything except the one variable factor of interest to him. That is totally different from the second activity of toxicity testing (although often miscalled 'toxicology') where one is simply trying to find out whether a chemical has any harmful effects. It is that sort of experimentation that I think Dr Festing is confusing with the type of refined animal experimentation where indeed precision, by the techniques that he mentioned and others, is important. You can see the same problem, I think, occurring in what has been said by Professor Peters and Dr Rees. When you are dealing with something very precise and controlled then you get these beautiful clean results; you know what you're looking for and you know how to assess it. Toxicity testing is not like that, and other sorts of, perhaps you will permit me to call them, more primitive medicine, are nearer to natural history. We are trying to find out something in the first instance. You can then develop from it, but please do not confuse the two and do not attack toxicology (which protects us all) for the wrong reasons.

Steven Rose (Open University): I speak as someone whose defence of animal experimentation has appeared in the 'New Statesman' headed (with a title the sub-editor clearly gave it): 'A Vivisector Defends His Trade'. I would like therefore to pick up from some of the points that Colin Blakemore made and I want also to pick up from the little announcement the Chairman made a moment ago regarding the removal of our name badges because I think this indicates the extent of the problem that we're facing. Why should we take our badges off if we leave this building at lunchtime? It seems to me that there is nothing to be ashamed of in making animal experiments for the purposes with which I suspect most of the people in this room who work with animals, like myself, are engaged. The less we actually have the courage of those convictions, the less we are prepared to defend those convictions and those actions openly, then the more likely we are to descend to a society which is going to become secretive of the work that research scientists, research clinicians and research doctors are doing. I think that would be dangerous. So my appeal is for openness. My appeal is also for not feeling so defensive about the need to respond to what the Animal Rights Movement is saying. The Animal Rights Movements' arguments are by and large a mixture of ignorance and cant and I think they need to be exposed as such directly. It is a great pity in some respects that in the course of this excellent day that has been organized, there isn't a more analytical approach to the sorts of arguments that the animal rights people are using, because I think we need to address them on their own ground.

The other point I want to make is one which follows from the issue of openness. I do think that one of the problems that we are facing is that we have been too defensive about what we're doing, too secretive and too closed in on ourselves. This applies to university research as well as to research fostered by the research councils and charities.

For example, the Open University is I think the only one in the country that now has an animal ethical committee in which animal experiments intended to be carried out in our own laboratories are assessed by a committee that contains both biologists and lay people from both inside and outside the university. And I think this openness is going to be essential if we are to move out from such a closed area to try and combat the ignorance which Colin Blakemore has identified. So I appeal for more openness and a more direct assault on some of the ignorance and the cant of the arguments of the animal rights people.

Donald Woods (Winchester): I wish to refer to the previous speaker's remarks about confrontation of the protagonists of animal rights with logical arguments. Certainly their arguments are mostly ignorance and cant.

A well-known antivivisectionist holds a post somewhat akin to 'Director of Religious Studies' at Essex University. He has written a book opposing animal experimentation on ethical grounds. He wears plastic shoes and doesn't use wool. I had the pleasure, on television, of getting him to admit that, if he were faced with the choice for his children of a medicine based on animal research, he would accept it. Yet, despite that admission, he still remains steadfast in his support of the antivivisectionists. Rational argument doesn't get very far in those circumstances.

Sir Walter Bodmer: I would like to make a general point on this topic. I think the important point about public understanding and reaching the public is not to convince the extreme antivivisectionists. You will never do that. The crucial thing is to answer their arguments so that the concerned public, starting from young school children to all the rest of us, do understand that there is a high moral ground in our stance. I don't think that one should worry about the extremists—you'll never change them. But, as many have said, we have to speak out in order to convince the public at large of the value of what we are doing.

Owen Wade (Birmingham University): I was the Dean of the Medical School of Birmingham at the time that the 1986 Act was being passed and we had a lot of trouble with antivivisectionists. As a consequence we threw open our labs to all the Members of Parliament in the West Midlands. Not all of them came of course, but quite a substantial number did. And I think that this was very helpful in some of the discussions at the time that the Act was passed.

The other thing that we did was to make a big effort to get the press to come and see our work. What I thought was very disappointing was that, although the individual reporters were quite convinced of the competence and the reliability, the ethical virtues and the importance of the work that was being done, this obviously wasn't the way their editors and their sub-editors saw it, because the headlines and much of the comment that was published was very antagonistic—presumably because the editors felt that this is what sells their newspapers. I have seen the same thing in some recent television programmes. So I think in a way, yes, it is the public that we need to reach, but I can't help feeling that it's partly the journalists. Perhaps the nation gets the quality of journalists it deserves and if that is so, well, I am very sorry.

Fiona Broughton-Pipkin (Nottingham): Continuing on the theme of educating the general public, I feel it is extremely important that we actually educate the school children. In this respect I was delighted to receive a few months ago notice of the establishment of a register of those involved primarily in scientific research of any sort, but of course this will include those of us who are involved with work using animals, who are willing to go and speak in schools. The Research Defence Society is, of course, one of the bodies which has supported this initiative and I would suggest very strongly that everyone who is involved ought, if they have not already done so, be placed on that register and be prepared to stand up and justify to the school children what we are doing.

5

An approach to the ethical issues posed by future animal experimentation

Professor Sydney Brenner FRS
Director, MRC Molecular Genetics Unit

Somebody said earlier that this meeting was ill-conceived because it is preaching to the converted. I disagree; it is preaching to the preachers—we have not yet got to the converted, let alone the heathen!

The defence of research unfortunately suggests a posture of justification and we now need to place much more emphasis on the positive value of our activities and also deal openly with the new questions that are raised continually by the rapid advances in biological science and technology. I will deal briefly with one such issue in this introduction; but first a few general remarks.

Some members of our audience have used the words 'vivisection' and 'antivivisection'. These may be thought to be old-fashioned and emotionally loaded but we need to talk about them because they say something about the public perception of animal experimentation. In very crude terms, the image still exists of a demonic scientist cutting up live cats, and causing pain and suffering in dumb animals. The question of pain was important historically, as we can see from the early laws on vivisection which happened to exempt cold-blooded animals from their purview and thereby embodied in statutes of the realm an ancient and mistaken physiological theory that these animals could not feel pain. Standards of experimental surgery have of course risen for animals—and indeed also for human beings—and I believe that the case for animal experimentation in many of the fields of biomedical research can be and has been easily justified by the benefits that it presents to the field of medicine. This is not so for the use of animals in toxicity testing. Much of what is done here is imposed by regulatory authorities in the interests of human welfare. As discussed by other speakers, it is here that scientists can make a positive contribution. As knowledge increases we can more readily devise tests that do not use living animals but have sufficient predictive value to allow at least an initial screening to be carried out. However, for this to have any impact the regulatory arms of government must also meet this challenge.

I would like now to consider some issues that arise from a new kind of experimentation which allows us to create, for example, transgenic animal models of human disease. We should accept from the outset that this technology is potentially explosive, because it could combine the fears the public have about genetic engineering with all of the issues of animal experimentation, and might also harness the growing public antagonism to science and technology in general.

We can begin with a very important ethical issue. We may think it a good thing to have something—perhaps we picked it up in Nature—but it does not follow that it would be good to bring that thing about by human intervention. Thus, there is a real difference between finding a mouse with inherited muscular dystrophy and creating one by direct genetic intervention.

In the first case, Nature, so to speak, bears the responsibility, but in the second, we do. Because it seems technically possible to produce genetic lines of animals with rheumatoid arthritis, cystic fibrosis and all sorts of genetic disease, we need to consider each case carefully. In the eyes of the public this will be seen as the very opposite of bringing about improvement by genetic manipulation and, unless we give a great deal of thought about how this is to be presented, we could come to be regarded as Frankensteins creating monsters. Consequently, we will evoke, what I believe is the real source of public anxiety, that mythic fear that scientists do 'unnatural' things.

Biomedical research has long followed a standard experimental style. Our methods are based largely on ablation followed by compensation. We found out how endocrine organs work by first removing them and then trying to replace them with extracts. We discovered vitamins by 'ablating' the diets of animals to generate vitamin deficiencies. In genetical analysis we 'ablate' genes by random mutations and, now, by directed mutation we propose to manufacture models of genetic diseases.

I believe in the future a new experimental approach will be adopted: we will seek answers to biological questions by composition rather than by decomposition. For example, when we approach very elaborate systems like the nervous system it may be of limited value to ask in how many different ways this can be broken either by extirpation or by genetic mutation. Rather, we may want to ask what evolution has added to a primitive fish brain to make a mammalian brain? We will want to put a fish gene into a mouse and ask whether it is expressed in the same way as its mouse counterpart and we will be asking questions such as: 'is the fish a proper subset of mouse?' This sounds strange today, but could become commonplace in the future.

One other important and constructive use of transgenesis is the creation of what can be called human surrogates. There is now good evidence that animal receptors are often different from their human counterparts. Thus, in the development of any new drug it is wise to start with the human receptor and this can be achieved by inserting its gene into an animal system; in many cases, of course, transgenic cells would suffice.

Transgenic experimentation is going to be central to future biomedical research and we should therefore be careful that we do not threaten it by ill-considered and possibly unnecessary experiments which may have a dramatic scientific impact but might also provoke a detrimental public response.

6

The discovery of drugs and animal research

Sir John Vane FRS
Chairman and Director, The William Harvey Research Institute

The intention of this article is not to plead the case for animal experimentation as a means of improving human and animal health, for this is now patently self-evident. Instead I shall sketch out some adventures in bioassay and drug discovery and show how these adventures have led, and will lead, directly to medical advances. In the process, I shall emphasize that it takes not months, not years, but decades to develop a discovery, as well as stressing the impossibility of predicting when a practical outcome will be achieved. I shall also give examples, partly from my own work, to illustrate not only the excitements of this type of research but also the internationalism of science.

The search for natural knowledge is the whole basis for achieving medical progress and it can lead in the most unexpected ways to important therapeutic advances. It is our ability to accumulate past experience in the forms of writings, art, scientific papers and verbal communication that sharply and uniquely distinguishes us as humans from the rest of the animal kingdom. It is our ability to learn from the experiences and experiments of others that is the essence of scientific understanding and progress.

Some years ago, I joined St Bartholomew's Hospital Medical College and founded there The William Harvey Research Institute. Interestingly, although William Harvey worked at Barts, they did not have a named Institute and since I intended working on the cardiovascular system, it was appropriate to name the Institute after the man who discovered the circulation of blood.

William Harvey also worked in Padua (from 1600 to 1602) where the famous Anatomy Lecture Theatre was sited. In those days, dissection of cadavers was illegal and so the students kept a watch-out. On the approach of police, the human body was hidden and replaced by the corpse of a pig. If the animal activists have their way, we shall be doing it the other way around!

It is difficult to quantify the enormity of knowledge and understanding that has stemmed from Harvey's discoveries. He experimented on many species, sometimes with great simplicity. For instance, he exposed the heart of a live snake (Figure 1) and noted that forceps, when placed on the great vein, led to a white, empty heart whereas, when placed on the arterial side, caused the heart to become engorged with blood. Imagine the excitement of noting this for the first time, together with the realization of what it meant. With the discovery of the circulation of blood, generation after generation of biologists have built a mountain of knowledge on its back. Think of the deficit, the deprivation, the dearth, the damage and suffering that would have ensued had this discovery not been made. A facile argument may be made that with modern imaging techniques, the circulation of blood in man could easily have been discovered, but those very techniques depend on keystones

Figure 1

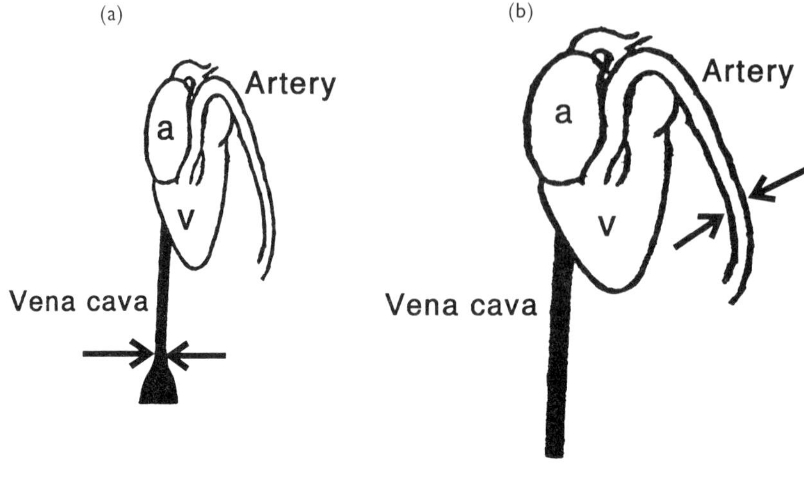

William Harvey's experiment on the heart of a snake. *Forceps placed on the vena cava (a) caused the heart to whiten and empty. Forceps placed on the artery (b) caused the heart to become engorged with blood.*

in the pyramid of knowledge built on Harvey's discovery, as well as on animal research, for their development.

I have illustrated an experiment performed on a whole animal and I cannot let the opportunity pass without asking the question: 'Does it upset you that scientists use snakes for their experiments?' And if the answer is 'No, after all, it's only a snake', just where do you draw the line? For my part, I am very comfortable with distinguishing between man on the one hand and all other species on the other. Some of the reasons for this I have given already, including our ability to accumulate knowledge, to communicate sensibly with each other in a sophisticated way and, most importantly, to contribute to the further progress of mankind and his animals. No doubt the animal liberationists, the animal rightists, the antivivisectionists would call me a speciesist. Incidentally, I always have difficulty in knowing how to describe these groups, varying as they do from the furtive and violent ones like the ALF to the more open ones such as BUAV and the NAVS. In preparing this article, I have tried to seek an appropriate single word definition for them. One that seems to me to be applicable comes from the shorter Oxford English Dictionary, as follows:

> Animalism—animal activity, sensuality; the doctrine which
> views men as mere animals; a merely sensual being.
> Animalist—one who takes the 'animal' side of a discussion;
> an adherent of animalism, a sensualist.

From now on I shall refer to opponents of animal experiments as 'animalists'. I am very clear in my own mind that we have to make strong distinctions between humans on the one hand and the beasts, birds and bugs of this planet on the other. We must get our priorities right and, for myself, I am more concerned with human rights than with animal rights. The plight of human victims of disease, rapists, muggers and bombers cries for our attention, as does that of millions of starving African children. And what is more, if I am buzzed by a mosquito, I take a swipe at it in the hope that I can prevent it from causing me unnecessary suffering. With the proviso that animal welfare is properly considered, I am comfortable in using other species as beasts of burden, for food, clothing or for animal experiments, and even more comfortable in actively destroying those which threaten man and his domestic animals, such as disease carrying vectors, vermin, parasites, bacteria and viruses.

Bioassay, the science of making quantitative biological comparisons, has its roots in ancient history when man began to use handy bits of his own body to make measurements. The foot, the hand, the palm and the digit are all self-explanatory. The cubit was the length of the arm from the elbow to the fingertips, whereas the fathom was the distance between the fingertips of the outstretched arms, interestingly approximating to the height. The definition I like best is the furlong which was the length of a furrow ploughed by a team of oxen between rests.

Clearly, these measurement would vary from person to person and this can be seen in the museum at Winchester where the yardstick (first based on the length of Henry I's arm) was changed from that measured against Henry VII to that for Elizabeth I. This type of biological variation has now been eliminated by making assessments against standard weights and measures. However, this does not eliminate another type of biological variation which becomes evident when man assesses himself against others in feats of strength or competitive athletic events. In using biological properties to make assessments, the variation clearly cannot be eliminated and so the need for reference standards of a biological nature sometimes remains, coupled with the need to make more than one assessment in order for the results to be more uniform and more reproducible.

Sir Henry Dale was the father of bioassay, and indeed the whole of modern physiology and pharmacology. In the early part of this century, when he was Research Director for the Wellcome Foundation, Dale was asked by Henry Wellcome to study the properties of the fungus, ergot, which grows on rye and makes it toxic. Out of this study came a veritable cornucopia of fundamental research, identifying chemical messengers of great importance to the physiology of the body. These included the amines acetylcholine, histamine and adrenaline. Indeed, it was the identification of acetylcholine as the chemical messenger in parts of our nervous system that led to Sir Henry Dale sharing the Nobel Prize in 1936.

Much of Dale's work relied upon the isolated organ bath, first invented by the famous Dutch pharmacologist, Rudolph Magnus (Figure 2). He found that if one took a small muscular piece of intestine or artery from an animal and bathed it with a warmed nutrient solution containing oxygen and glucose and the salts of the blood, the muscle of the preparation would stay alive for hours if not days. When the length of the muscle was recorded by a simple lever system, the addition of different chemicals or drugs would make the muscle contract or relax and this could be recorded on a smoked piece of paper wrapped around a very slowly revolving drum. The extent of the contraction depended on the dose of the chemical introduced into the solution and this provided a remarkably simple and effective way of testing different solutions for their potency or activity. The only limitation to this

Figure 2

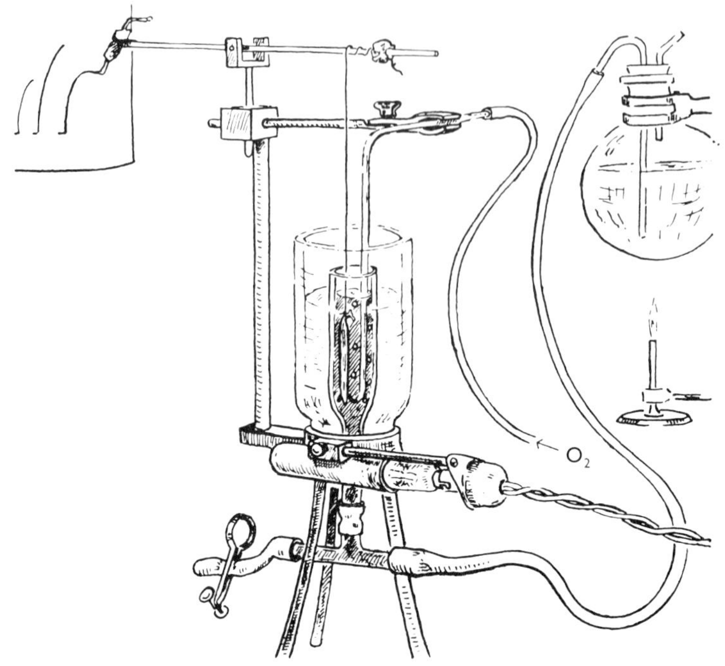

The isolated organ bath designed by Rudolph Magnus. *The piece of tissue containing smooth muscle is suspended in an oxygenated physiological salt solution in the inner chamber. The surrounding chamber contains water maintained at 30–37°C by means of the electric heating element. Contractions of the smooth muscle cause the lever to rise, the extent of contraction being recorded on smoked paper covering the rotating drum. From Burn (1952) with permission.*

type of technique is that the piece of tissue used has to be thin enough to be kept alive by the diffusion of oxygen and glucose into the tissue from the bath.

One of Dale's students was my own mentor, Professor Harold Burn. He introduced me to methods of bioassay and, in particular, to one where the tissue was too bulky to be kept alive in an isolated organ bath, so that a nutrient solution containing oxygen was perfused through its blood vessels as shown in Figure 3. Again, this isolated tissue stayed alive *in vitro* for hours and provided much information on substances which contracted or relaxed the vessels.

It was within a most fertile atmosphere of powerful intellects that my own scientific training was shaped. Others who actively influenced me at the time were those notable scientists who came here from Germany, including Feldberg, Vogt, Bulbring and Schild. I came back to the U.K. from a post-doc in Yale to work at the Royal College of Surgeons with Bill Paton and in those happy days developed a bioassay technique of my very own. This involved taking a stomach from a rat (Figure 4), using the fundal portion and cutting it into a zig-zag strip and hanging it in an isolated organ bath, recording its length with a lever. This rat stomach strip was extremely sensitive to

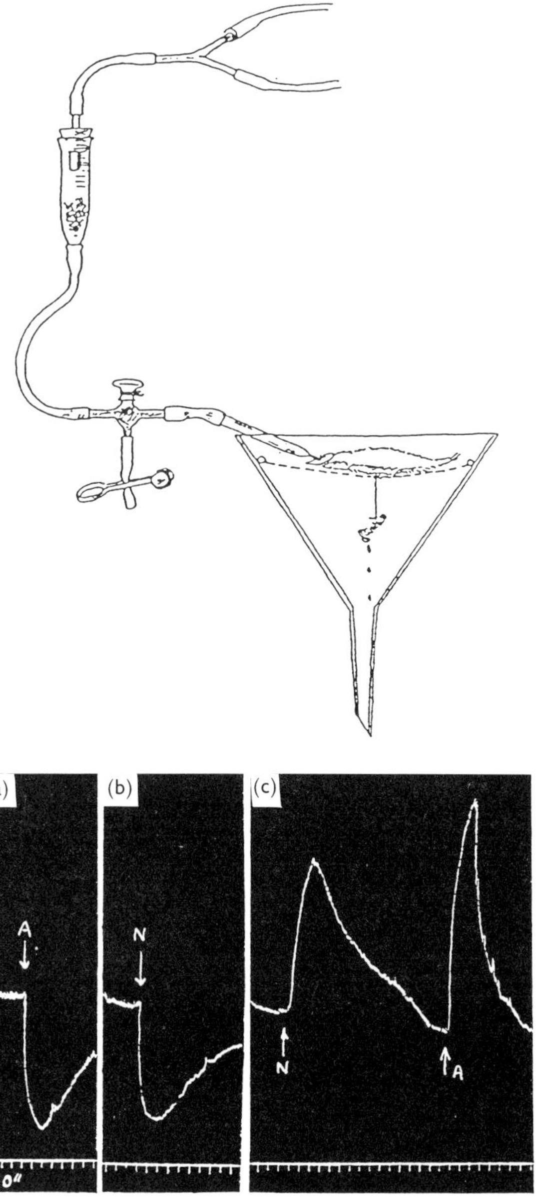

The perfusion of a whole organ with physiological salt solution. *The isolated rabbit ear (placed in a funnel) is perfused through the central ear artery with aerated, warmed physiological salt solution (top). The outflow from the veins is measured by a flow recorder (bottom). The record shows a decrease in flow (vasoconstriction) caused by adrenaline (A) and noradrenaline (N) (a, b). After blockade of α-adrenoceptors by benzylimidazoline, adrenaline and noradrenaline produced vasodilatation (c). From Burn (1952) with permission.*

Figure 4

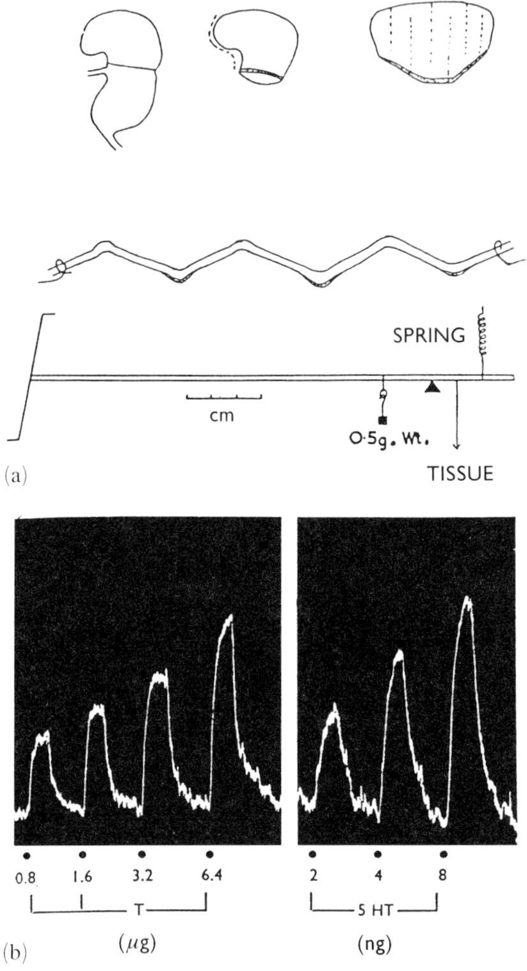

Bioassay to assess the effect of chemical messengers on rat fundic strip. *(a) Preparation of the rat fundic strip. (b) Use of the rat fundic strip to compare the activity of 5-hydroxytryptamine (5HT) against tryptamine (T). From Vane (1957).*

another chemical messenger 5-hydroxytryptamine (5HT). Figure 4 also shows the bioassay of tryptamine itself, from which the more potent 5HT is derived, against increasing doses of 5HT in an isolated organ bath. These are measured in nanograms which are thousandths of a millionth of a gram; thus, it is clear that the assay is very sensitive and, at the time of its development, much more sensitive than any assays depending upon chemical measurement. This indeed is an important aspect of bioassay, especially when newly identified endogenous substances are being investigated.

Another of Dale's students, Jack Gaddum, also made vital contributions to the development of bioassay. He made the results more meaningful by introducing a statistical approach and also introduced a third basic tech-

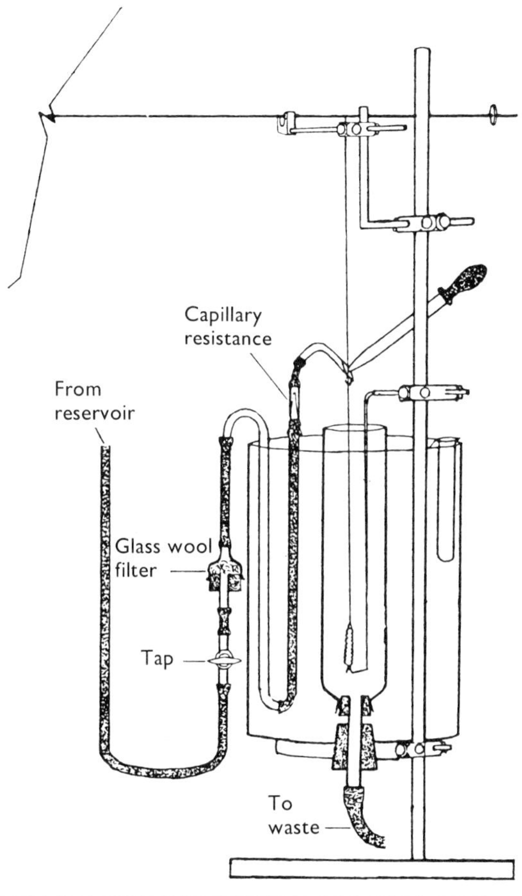

The technique of superfusion of isolated organs (Gaddum, 1953). *Warmed, aerated physiological salt solution is allowed to trickle over the tissue suspended in the inner bath. Drugs are added by means of the pipette.*

nique, that of superfusion. In this case, rather than use an isolated organ bath full of a nutrient solution, he dribbled a continuous stream of fluid over the outside of the isolated piece of tissue in order to keep it alive. The great advantage of this technique was that it reduced the volume of fluid involved, thereby making the assay much more sensitive (Figure 5). Gaddum said:

> 'The pharmacologist has been a "Jack of All Trades", borrow-
> ing from physiology, biochemistry, pathology, microbiology
> and statistics but he has developed one technique of his own
> and that is the technique of bioassay'.

I then began to develop Gaddum's superfusion system. If a single tissue can be superfused, why not more? Figure 6 shows an experiment using three different tissues, chosen for their sensitivities to different substances. Illustrated are the effects of some of the prostaglandin family including thromboxane A_2, prostaglandin E_2 and another type of prostaglandin. Different tissues from different parts of the body have different sensitivities to

Figure 6

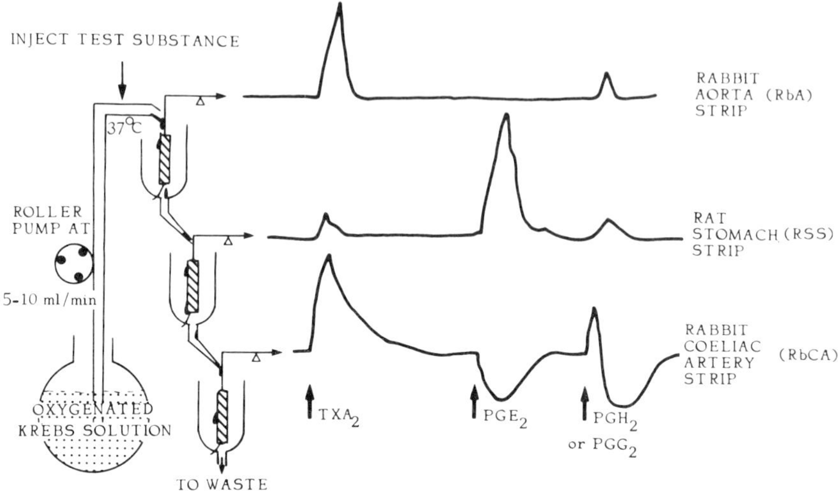

The technique of cascade superfusion. *Warmed, oxygenated, physiological salt solution is allowed to trickle sequentially over a bank of tissues (three in this case). This experiment demonstrates the formation of a novel prostaglandin, that relaxes rabbit coeliac artery, from the endoperoxide precursors PGG_2 and PGH_2. Abbreviations used: PG, prostaglandin; TXA_2, thromboxane A_2.*

different chemical messengers. By carefully choosing different tissues, selectivity can be added to the sensitivity of bioassay so that a substance which caused contraction of, for instance the strips of rabbit aorta and the strips of rabbit coeliac artery, was almost certain to be thromboxane A_2.

Over a period of some years, we built up a whole battery of different tissues whose sensitivities to various substances was defined. The number of tissues used was by no means confined to those shown in Figure 7. Indeed, it was said that every new graduate student in my laboratory had to find his own bioassay tissue which uniquely measured, either by itself or in combination with others, a particular chemical messenger or hormone. Take any part of the body and you can be sure that the pharmacologist has excised it, hung it in an isolated organ bath and measured changes in its length produced by different stimuli.

Let us now look at some instances where the techniques of bioassay have initiated exciting progress in medicine. One of Dale's discoveries was the chemical messenger histamine, involved in the triple response, in anaphylaxis and, more importantly, in physiology as the natural stimulus for acid secretion by the stomach. In the very early days, Dale referred to 'extrinsic' and 'intrinsic' histamine to define two separate activities, and when the well known antihistamines were discovered through animal experiments in the 1940s it was soon recognized that these drugs would prevent some actions of histamine including those involved in hay fever, but would not antagonize others including the secretion of acid by the stomach. It was not until the 1960s that a second class of histamine receptors (called H_2 receptors) was defined through the work of Schild and Black and others, giving Jim Black the opportunity to seek his second major therapeutic inter-

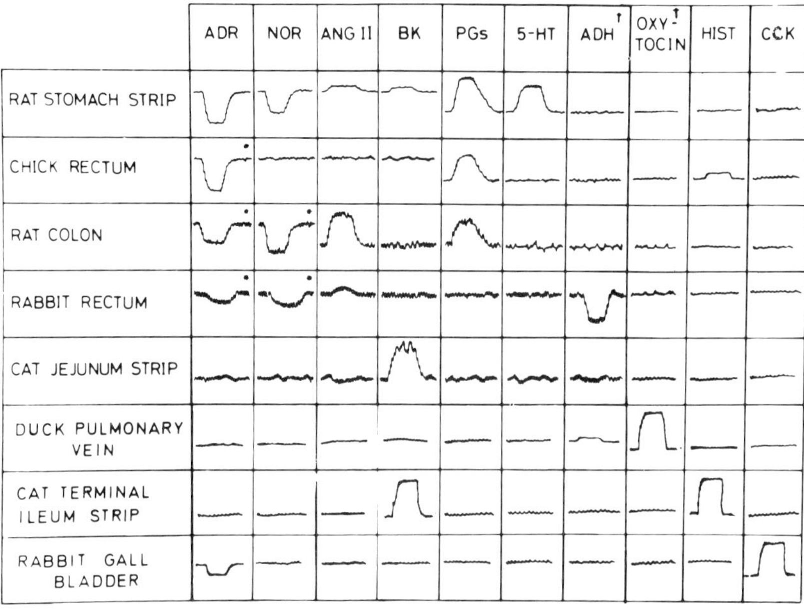

The sensitivities of different tissues to some endogenous chemical messengers. *By using the right combination of assay tissues, virtually any mixture of substances can be assayed. Abbreviations used: ADH, antidiuretic hormone; ADR, adrenaline; ANG II, angiotensin II; BK, bradykinin; CCK, cholecystokinin; HIST, histamine; NOR, noradrenaline; PGs, prostaglandins.*

vention. He had already become famous for his development of the β-blockers, still amongst the world's most used drugs, whilst he was at ICI. After moving to Smith-Kline and French with Bill Duncan, together with a small but enthusiastic team, they tackled the problem of H_2 antagonists.

Jim Black is renowned for his ability to bring accurate measurement into pharmacology and this was no exception. He measured histamine-stimulated acid secretion in rat stomachs in the isolated organ bath and the effects of histamine in pieces of isolated heart tissue which also contained H_2 receptors. By this second piece of stunning scientific development, under the inspired direction of Bill Duncan, the team produced cimetidine or Tagamet for the treatment of peptic ulcers, as the first of a new series of H_2 antagonists destined to become another class of blockbusting drugs. Ranitidine, produced by Glaxo, was the second major drug in this field. For his astonishing insight into pharmacological receptors, in large part using bioassay techniques, and for his contributions to pharmacotherapy, Sir James Black shared the Nobel Prize in Physiology or Medicine in 1988 with George Hitchings and Trudy Elion. George is now a sprightly 86 year old and Figure 8 shows him in 1960 with, amongst others, Trudy Elion, Roy Calne and Joe Murray on the steps of Harvard Medical School. The other stars in the picture are Tweedledee, Tweedledum, Titus and Lollipop who had all received transplanted kidneys under the cover of the then new drug Imuran or azathioprine, invented by Hitchings and Elion. The work on these dogs led directly to transplantation of organs between humans and now, for kidneys alone,

Figure 8

Joe Murray (extreme right), George Hitchings (centre), Trudy Elion and (on her right) Roy Calne amongst others on the steps of Harvard Medical School (1960). *Also present are Tweedledum, Tweedledee, Titus and Lollipop who received successful kidney transplants through the use of the drug azathioprine.*

10 000 a year are taking place. In 1990, Joe Murray and Don Thomas were awarded the Nobel prize in Physiology or Medicine for their work on organ transplantation.

The benefits of the clutch of novel drugs invented by Hitchings, Elion and Black — β-blockers, H_2 antagonists, antibacterials, antimalarials, gout-suppressants and immunosuppressants — have been the saving of millions of people from death and the substantial improvement in the quality of life of tens of millions of others. The cost in animals used for research was far lower than the cost of those eaten, collectively, by persons attending the average meeting. The cost : benefit ratio in terms of suffering is easily calculable, wholly verifiable and overwhelmingly acceptable to all those concerned with the improvement of the human condition.

My next example arises from the discovery of 5HT. In the early days, as with so many biological activities, it was known by several different names such as serotonin and enteramine and it was only in the period after the 1940s that these various activities were all identified as 5HT which was shown to be a constituent of several different tissues. In the 1960s different receptors for this amine were discovered and this led yet another inspired group of receptor pharmacologists at Glaxo, under the brilliant leadership of David Jack, to begin to analyse the control of those blood vessels in the head which were thought to be implicated in a migraine attack. The current theory was that dilatation of these blood vessels led to the pulsating headache: in other words, that migraine was a malfunction of blood vessels of the head, which could be remedied by selective constriction of the dilated vessels.

Pat Humphrey led this work and showed that 5HT constricted some of these blood vessels through an activity on a special 5HT receptor. After much laborious analysis of receptor types (a characteristic of David Jack's approach to the discovery of new drugs) they found that strips of dog saphenous vein contained the same special 5HT receptors as the blood vessels within the head and so they began to use this simple preparation as a screening test. The aim was to come up with a long-lasting substance which would stimulate these receptors in the same way as 5HT, but selectively, without stimulating all the other 5HT receptor types.

It showed great courage and faith in bioassay for a Research Director to spend millions of pounds on developing a new substance using a simple bioassay such as the dog saphenous vein, but he had the confidence that it would predict substances of value for the alleviation of migraine. To illustrate the importance of bioassay I have, of course, made a gross over-simplification. No compound could progress from an *in vitro* assay to man without careful and painstaking experiments *in vivo*—in anaesthetized dogs, for example—to ensure that the required constriction of blood vessels in the head was selective, potent and achievable after oral administration. Extensive safety studies in at least two species are also required to make sure that predictable toxicity has been avoided.

All this takes time and in 1984, some 12 years after the project was initiated, sumatriptan was identified as the best compound for alleviation of migraine. It was developed as rapidly as possible but as with all new drugs this took many more years. This year, sumatriptan is about to be marketed around the world for the alleviation of migraine, having been shown in careful clinical trials to have a strongly positive effect.

So far my illustrations have relied on either imitating or antagonizing amines as chemical messengers, but now I want to discuss the second group of chemicals known as peptides. These substances consist of amino acids joined together in a chain containing from two to several hundred (the larger ones being called proteins). One of the first peptide system to be intensively studied was the renin–angiotensin system. In the 1950s it was shown that an enzyme called renin was secreted by the kidney into the bloodstream where it cleaved a decapeptide called angiotensin I from a plasma protein. This decapeptide was inactive but a second enzyme known as angiotensin-converting enzyme (ACE) chopped two more amino acids from its end changing it into the potent pressor octapeptide, angiotensin II. At the time, it was suspected that the renin–angiotensin system was of importance in cases of high blood pressure, meaning that too much angiotensin II may bring about hypertension, but there was no experimental evidence to link the two together.

The story which follows displays several facets of biological discovery. Certainly, it illustrates the internationalism of science which I have already mentioned, but there are other well-defined elements including unpredictable outcomes of research, opportunism, serendipity, coincidence and good fortune.

In 1964, Sergio Ferreira came from Brazil to my laboratories at the Royal College of Surgeons. In Brazil, he had studied under Rocha é Silva who discovered bradykinin, another important peptide which he first isolated from the poisonous venom of the Brazilian snake, *Bothrops jararaca*. Working on the old principle that venoms sometimes contained not only noxious substances but also others which potentiate their effects, Sergio Ferreira isolated in 1965 from the venom of the same snake, a factor which he called bradykinin-potentiating factor (BPF). He came to us as a post-doc, carrying some BPF in his pocket, choosing me to work with (as he later confessed) not because of my reputation but because my laboratories were close to the London School of Economics where his wife wanted to take her PhD. He himself would have preferred to go to Oxford.

At the time, we were working on the renin–angiotensin system and I suggested to Sergio that he should test his snake venom extracts on the various enzymes involved. However, he had his own ideas and within a week or so he had persuaded me to work on bradykinin rather than he working on angiotensin! It was only two years later than another of my colleagues, Mick Bakhle, agreed to test the snake venom extract on ACE and showed it to cause a potent inhibition.

We followed this up on various bioassay preparations and also in the whole animal. We had already shown that the lungs contained high levels of ACE, using a simple perfused lung preparation in which the perfusate super-fused a rat colon highly sensitive to the contractor activities of angiotensin. It was in this and similar preparations that we also showed that BPF inhibited the conversion of angiotensin I to angiotensin II.

At the time, I was a consultant to the drug company Squibb in New Jersey and I went to them with the proposal that they should take up the study of this snake venom extract which we then knew to be a mixture of peptides. By using one of the purified peptides from it, they would be able to prove or disprove the concept that angiotensin was important in high blood pressure. Indeed, in parallel work when he returned to Brazil (now con-vinced that ACE inhibition was important) Ferreira showed that BPF reversed the rise in blood pressure in rats which was associated with a massive release of renin when the blood supply to a kidney was restored after being clamped for 6 h (Figure 9). It also worked in other experimental models of hypertension. I visited Squibb three times a year and each time found that their initial enthusiasm for the project was waning, for their marketing people did not comprehend that proving a concept with an extract of a snake venom could possibly lead to a new drug. Peptides have to be injected rather than given orally and they reiterated that they could not sell an antihypertensive drug which had to be injected. They were in the

Figure 9

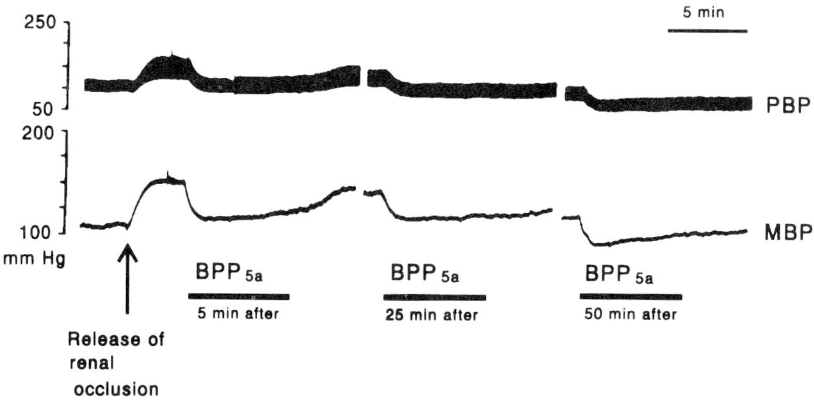

Inhibition by bradykinin-potentiating peptide (BPP$_{5a}$) of the rise in blood pressure following release of occlusion of the renal artery in the rat. *Abbreviations used: MBP, mean blood pressure; PBP, pulsatile blood pressure. Reproduced from Krieger et al. (1971) with permission.*

business of selling drugs and not of proving concepts. Nevertheless, my two main contacts at Squibb, the vice president in charge of research, Arnold Welch (with whom I spent my post-doc when he was a Professor of Pharmacology at Yale), and his deputy Chuck Smith remained enthusiastic; this eventually led to the synthesis of 1 kg of teprotide, the most active peptide from BPF. This indeed lowered blood pressure in hypertensive man, thus proving the concept that inhibition of converting enzyme was an important new therapeutic target.

In 1973 I joined the Wellcome Foundation, one of Squibb's competitors and was no longer able to consult with them. It was within the next two or three years that Ondetti and Cushman at Squibb solved the problem of changing a peptide inhibitor of ACE into a non-peptide inhibitor with oral activity. This eventually led to captopril, the first ACE inhibitor to be marketed as an antihypertensive drug. This has also become a blockbuster drug of substantial importance, as have other ACE inhibitors since marketed by other companies. Incidentally, but importantly, ACE inhibitors have also found other uses in man, such as in the treatment of heart failure.

What a tortuous pathway this history involves. All of the elements of basic research—unexpected discoveries, unexpected outcomes, internationalism in science and fortuitous alliances—are there, as well as the now well-established fact that the marketing people in a drug company consistently fail to predict the usefulness of drugs (such as propranolol, cimetidine and captopril) that involve important new concepts, and therefore actively oppose their development.

My next illustration comes from a group of yet another family of chemical messengers, the prostaglandins. At the Royal College of Surgeons, we first entered this field in the mid 1960s and, by the use of bioassay techniques, helped to elucidate the pharmacology of these substances. It was during this period that Richard Gryglewski came from Poland for the first of his many visits to my laboratory. He is the Professor of Pharmacology in Cracow and an ardent supporter of Solidarity, under whose banner he was successfully elected to the Rectorship of Cracow University. Our collaboration over the past 25 years on the pharmacology of the prostaglandins and other mediators has been a most fruitful one which continues through visits for a few months each year by Richard to The William Harvey Research Institute as well as through visits by the young members of his staff who come to supplement their training. I also visit his laboratory in Poland, as do some of our young people.

The prostaglandins were first isolated in seminal fluid in 1933. Von Euler, who was later to share the Nobel prize for other seminal contributions in the field of the adrenergic nervous system, called them prostaglandins because he believed erroneously that they came from the prostate gland.

In the 1960s there was great interest in these potent acid lipids and we studied in some depth their pharmacology, especially their activity in causing blood vessels to dilate. We also helped to delineate their release from and inactivation by different organs of the body.

At the time, we were also working on the type of anaphylaxis or shock which underlies the pathology of asthma. This involved studying the chemical mediators or messengers released during anaphylactic shock from lungs isolated from the guinea-pig. There were so many chemicals released that we had to increase our bioassay cascade to include six different tissues, each one selectively printing out in a dynamic way the release of one of these

Figure 10

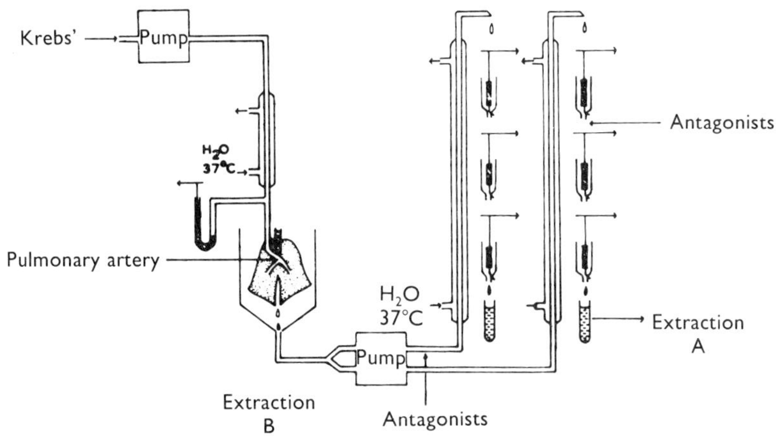

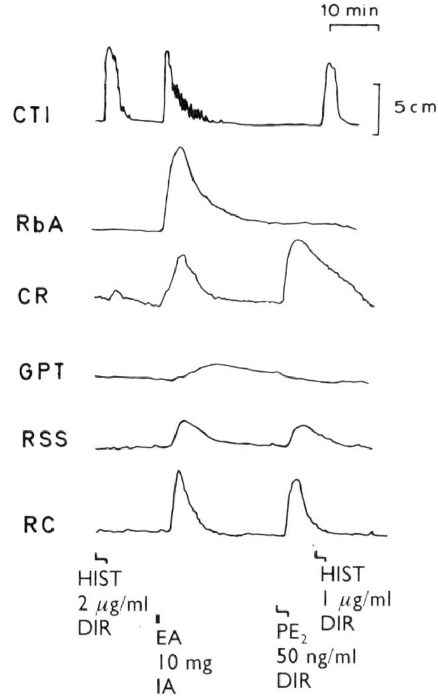

The detection of substances released by challenge of sensitized (with egg albumen), perfused guinea-pig lungs. *A set of six tissues was necessary to detect the various substances released during anaphylactic shock. Histamine (HIST) and prostaglandin E_2 (PE_2) were administered directly over the tissues (DIR), while egg albumen (EA) was given through the lungs (IA). Abbreviations used: CR, chick rectum; CTI, cat terminal ileum; GPT, segment of guinea-pig trachea; RbA, rabbit aortic strip; RC, rat colon; RSS, rat stomach strip. From Vane (1972), with permission from S. Karger AG, Basel.*

substances from the lungs (Figure 10). We caused anaphylactic shock by injecting egg albumen, a protein to which the guinea-pig had previously been sensitized. We showed the release of different mediators, some of which were well known, such as histamine, and slow-reacting substances of anaphylaxis, characterized later in 1979 by Samuelsson as a mixture of products of arachidonic acid which he re-named 'leukotrienes'. We also showed that prostaglandins E_2 and F_{2a} and a newly found substance which we called rabbit aorta contracting substance (RCS) were released during this pathological reaction. When Priscilla Piper came to work with me as a PhD student from Harry Collier's laboratory, she brought with her his interest in aspirin and similar drugs. We went on to show that mefenamate and other aspirin-like drugs selectively abolished the release of RCS.

It was this work which led me directly to the idea that aspirin could inhibit the natural production of prostaglandins, thereby explaining its therapeutic effects. I well remember formulating this concept over the weekend whilst I was writing a review. On the Monday morning, I went into the laboratory and said to Priscilla Piper, Sergio Ferreira and others that I thought I knew how aspirin worked. This was before carrying out the actual experiment! On that Monday, I took the very tissue which had led to the concept—the guinea-pig lung—homogenized it and precipitated by centrifugation the cell debris, leaving a soup of the cell contents which I knew from the work of Anggard and Samuelsson would contain the enzyme that synthesizes prostaglandins from arachidonic acid. With a little of this enzyme in different test tubes, I added the precursor arachidonic acid and then put in different doses of aspirin, salicylic acid, morphine or indomethacin. I included morphine because it was an analgesic working by a different mechanism. I measured prostaglandin formation by bioassay. By the end of that day I was convinced that aspirin (but not morphine) indeed inhibited the biosynthesis of the prostaglandins. After a further 3 weeks work in which the experiment was repeated in several different ways, the evidence was compelling (Figure 11).

Two new PhD students joined in this work—Rod Flower who had been with us previously as a technician before taking first class honours in Sheffield and Salvador Moncada who came to do a PhD from Honduras. They, together with Sergio Ferreira, over the next few years developed the concept into the now generally accepted theory that the mechanism of action of aspirin-like drugs is through inhibition of prostaglandin biosynthesis. Clearly, this research on the biochemical events which explain the way in which aspirin works has had enormous implications. New aspirin-like drugs have been discovered and marketed and new uses for aspirin have emerged, including its use to prevent heart attacks and strokes.

Another illustration of the internationalism of science was the way in which two leading groups, that headed by myself and that headed by Bengt Samuelsson in Stockholm, were able to leapfrog each other in scaling the mountain of knowledge that was being developed in this field. Earlier, as I have mentioned, we had discovered a substance which we called rabbit aorta contracting substance and in 1974 Bengt Samuelsson characterized this chemically and renamed it 'thromboxane A_2'. It was in looking for the enzyme that generates thromboxane A_2 from arachidonic acid that we came across, by serendipity, yet another previously unknown substance.

Thromboxane A_2 is generated by platelets and acts as a powerful vasoconstrictor substance as well as causing platelets to clump together. At

Figure 11

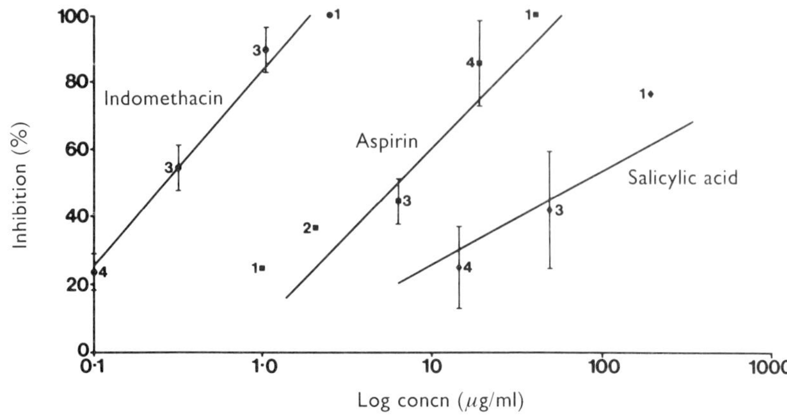

The inhibition of prostaglandin synthesis by aspirin-like drugs.
Arachidonic acid was incubated with the supernatant from a guinea-pig lung homogenate in the absence of drug, or in the presence of various concentrations of nonsteroidal anti-inflammatory drugs. The ability of the drugs to inhibit the synthesis of prostaglandins generally ranked with their effectiveness as anti-rheumatic agents. Figures represent numbers of observations. From Vane (1971), reprinted with permission from Nature, copyright 1971.

this point, I had moved to Wellcome with Salvador Moncada who headed our prostaglandin research department. We began a search, using our classic techniques of bioassay, for tissues other than platelets that could make thromboxane A$_2$. We took extracts from kidney, lung, spleen and aorta and incubated them with the prostaglandin precursors to find out by bioassay whether the extracts made thromboxane A$_2$. Kidney and lung did not generate RCS or thromboxane A$_2$ but did make prostaglandins, whereas the spleen made both. However, the aorta did not seem to make either, although the precursor disappeared. This meant either that the tissue simply destroyed the precursor or that our bioassay techniques were not picking up an active product; the latter turned out to be the case. When we turned to other bioassay tissues such as the rabbit coeliac artery (Figure 12) we found that the aortic extract was indeed manufacturing a dilator substance which did not correspond to anything then recognized. Very soon afterwards, we also found that PGX, as we called it, inhibited the clumping of platelets. This work led to the identification in 1976 of a newly discovered prostaglandin. We renamed it 'prostacyclin' and, in collaboration with another pharmaceutical company, the Upjohn Company in Kalamazoo, we characterized its chemical structure and synthesized it.

Prostacyclin is chemically unstable, disappearing within a few minutes. It is a powerful vasodilator and potently prevents platelets from clumping together. It is made by the endothelial cells which line the blood vessels and helps to keep them clean, preventing platelets from sticking to them to form clumps and thereafter thrombosis. We developed prostacyclin as a drug for use in overcoming obstructive vascular disease. Its instability was surmounted by making it up in special alkaline solutions and it was tested by intravenous injection over several days in different kinds of peripheral vascular disease. Sadly, within Wellcome, prostacyclin became a

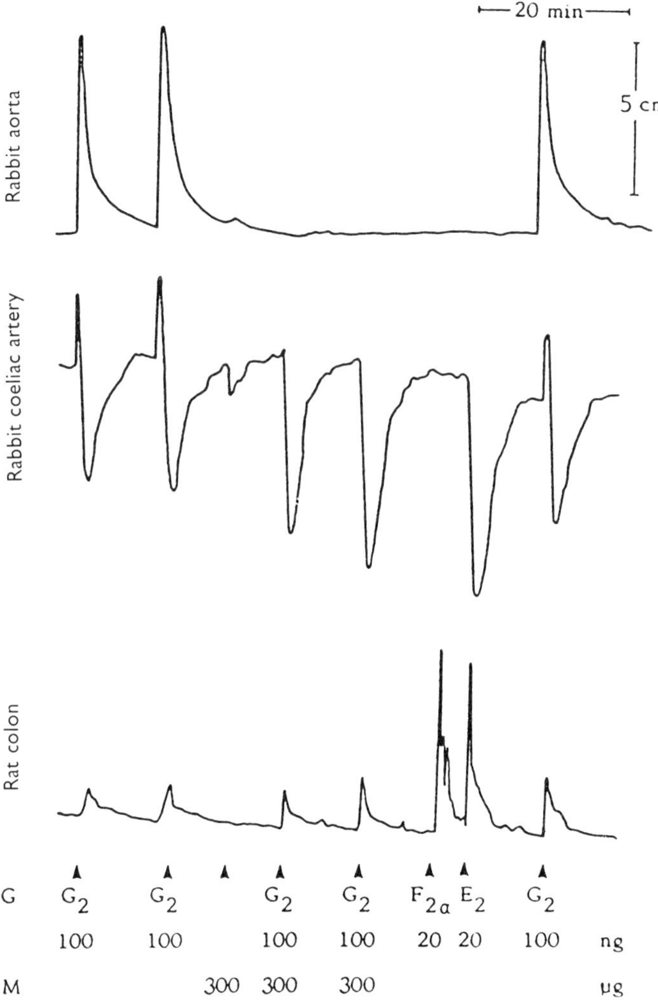

The synthesis of PGX (later PGI$_2$) by aortic microsomes. *The prostaglandin precursor PGG$_2$ (G$_2$) contracted rabbit aorta, caused a brief contraction followed by a relaxation of the coeliac artery and gave a small contraction of the rat colon. In contrast, when PGG$_2$ was incubated for 1 min at 22°C with pig aortic microsomes (AM) a novel product was formed (PGI$_2$) which did not contract the rabbit aorta or the rabbit coeliac artery but caused a profound relaxation of the rabbit coeliac artery. From Bunting et al. (1976) with permission.*

political football because of its close association with myself, and the internecine strife that was generated prevented its full development even though it earns a few million pounds each year. The overall lack of vision of the marketing side of a pharmaceutical company was once more demonstrated. Even the medical division failed to understand the importance of the discovery. However, other drug companies, including Ono in Japan and Schering in Germany, picked up the baton. Schering have now developed a stable analogue called iloprost which is about to be marketed for the treatment of obstructive diseases of the circulation for which there is no other effective treatment. As with the first prototype ACE inhibitors, orally active compounds are also on their way. Again, the time scale is inordinately protracted with two decades passing between discovery and the marketing of a compound active by mouth.

My last two examples both come from work on the endothelial cell, that monolayer of cells which lines the blood vessels of our circulation (Figure 13). Initially, they were thought to act as a sort of dialysis membrane allowing nutrients from the blood stream to diffuse through to the muscles of the arteries underneath without letting proteins or blood cells escape. Now, we know that these marvellously sophisticated cells are virtually in control of the blood circulation.

Within the past few years, it has been recognized that these cells produce not only vasodilator compounds such as prostacyclin, but also vasoconstrictors—compounds that will narrow the blood vessels. One such compound was known to be a peptide and, using the tools of molecular biology, a Japanese group under the leadership of Tomoh Masaki was able to characterize this vasoconstrictor peptide as having 21 amino acids: they called it endothelin (Figure 14). The paper was published in *Nature* on 31st March, 1988 and it is an extraordinary example of the completeness of a piece of biochemical and biological work. In fact, this six-page paper is based on the PhD thesis of Masashi Yanagisawa. The long-lasting hypertensive activity of endothelin is shown in Figure 15. I visited Tokyo the week after the paper was published and arranged to meet with Masaki and Yanagisawa who kindly gave me a sample of the then precious endothelin. We rapidly developed the pharmacology of endothelin as did the Japanese and many others over the next 2 years. We now know that endothelin is more potent on venous than on arterial smooth muscle and that it is avidly removed by the pulmonary circulation. Endothelin is the most potent constrictor of vessels (leading to a

Figure 13

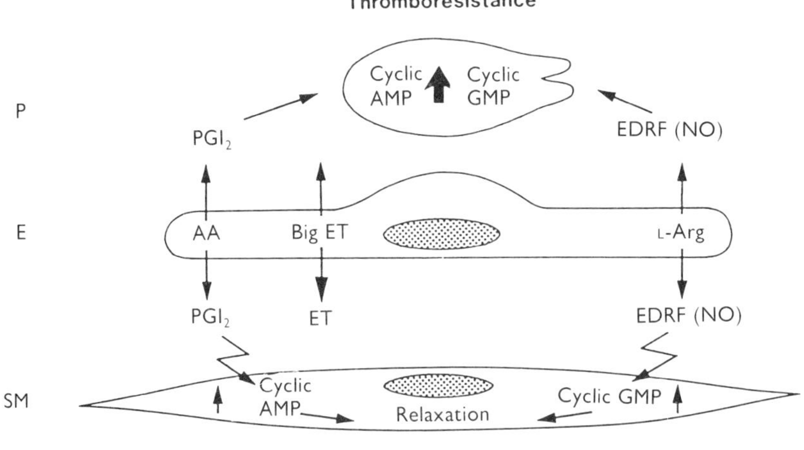

The vascular endothelial cell (E) and how it may affect the circulating platelet (P) and the underlying smooth muscle cells (SM). *Abbreviations used: AA, arachidonic acid; EDRF, endothelium-derived relaxing factor; ET, endothelium; NO, nitric oxide.*

rise in blood pressure) yet discovered, easily beating the previous record holder angiotensin II by some 10-fold. Clearly, this chemical messenger is yet another potential starting point for an antihypertensive drug, based either on preventing its formation by the endothelin-converting enzyme or on preventing its activity with a receptor antagonist (Figure 14). Many drug companies are now looking for antagonists of its action or of its production. Surely within the next 5 years the concept of preventing the actions or release of endothelin to reduce hypertension will have been tested. Certainly nature would not have implanted the production of such a potent substance into the endothelial cell without purpose, although raising the blood pressure may not be its only function.

My last study does not involve a peptide, an amine or a prostaglandin but a far simpler molecule. Although now at the leading edge of research, it also turns out to have a long lineage.

For more than 100 years substances like amyl nitrite and nitro-glycerine have been used to cause a widening of the arteries in conditions such as angina. Indeed, Alfred Nobel who made his fortune out of dynamite (which was a simple but stable pharmaceutical preparation of kieselguhr mixed with nitroglycerine) wrote to a friend saying: 'It sounds like the irony of fate that I should be ordered by my doctor to take nitroglycerine internally'.

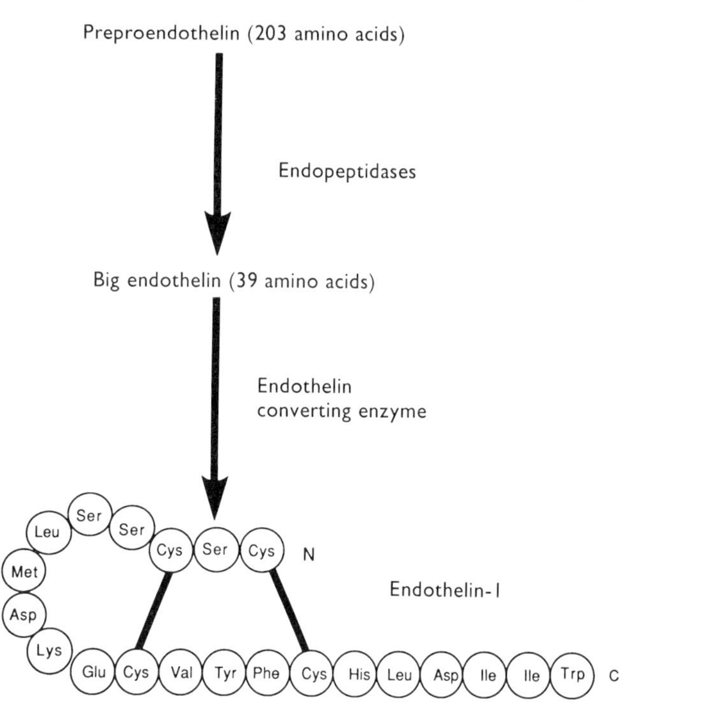

Figure 14

The proposed pathway of formation of endothelin-1 and its amino acid composition.

Figure 15

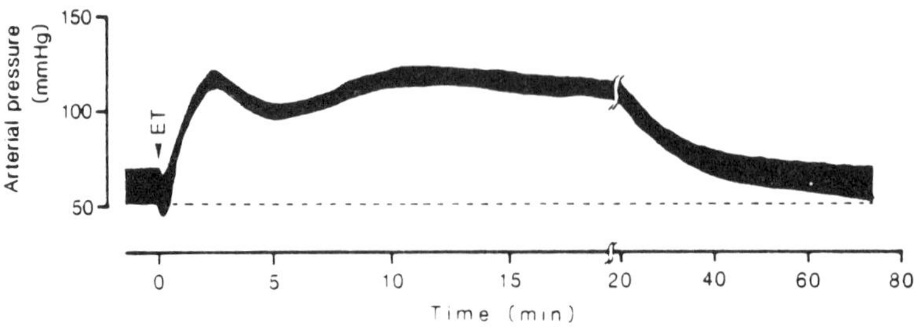

The effect of endothelin-1 (ET) on rat blood pressure after a single dose. *The blood pressure is increased for more than 60 min. Reproduced from Yanagisawa* et al. *(1988) with permission. Copyright Nature, 1988.*

Work in the 1970s showed that this type of nitrate was converted in muscular cells to the simple molecule nitric oxide (NO) and it was this substance that brought about the relaxation of arterial and venous muscle.

Coming from an entirely different aspect and using, as a bioassay tissue, strips of muscle cut from rabbit aorta, Furchgott in 1980 showed that after these strips were contracted with noradrenaline, acetylcholine caused a relaxation (Figure 16). However, when the endothelial cells were rubbed off the strips, acetylcholine then caused a contraction rather than the relaxation. Furchgott could not escape the conclusion that when the endothelial cells normally attached to the vascular strip were stimulated by acetylcholine, they released a substance which dilated the associated muscle: he called it endothelium-derived relaxing factor (EDRF).

In the next few years EDRF was shown to be very unstable, a powerful vasodilator and to have important inhibitory functions on platelets. Many research laboratories looked to see whether it was an amine, a peptide or a prostaglandin, without much success. After I left Wellcome in 1985, Salvador Moncada became Research Director and he continued the search for the chemical nature of EDRF. To do this, he grew endothelial cells on microscopic beads and put some 20 million cells into a small column at the top of a cascade of rabbit aortae (Figure 17). When he stimulated the column with an EDRF-releasing substance, the rabbit aortae relaxed in a fashion characteristic for EDRF. This and other work led to his identification of EDRF as NO.

For 50 years, we pharmacologists have been discovering an extraordinary number of potent and important chemical messengers many of which I have already mentioned. These include amines, peptides and prostaglandins, all complex organic molecules. In 1987, one of the most potent chemical messengers yet discovered turned out to be a simple one: one combination of the two main elements of the air—nitrogen and oxygen.

As with many hormones, we learn most about their function when we have access to substances that antagonize their action or production. With the prostaglandins, for example, the discovery that aspirin and similar drugs prevented their production gave an enormous boost to our understanding of their functions. The same has happened in the field of NO research. We now know that NO is formed enzymically in the body from the amino acid L-arginine (Figure 18). Furthermore, a simple arginine analogue, methylarginine, interferes with the generation of nitric oxide.

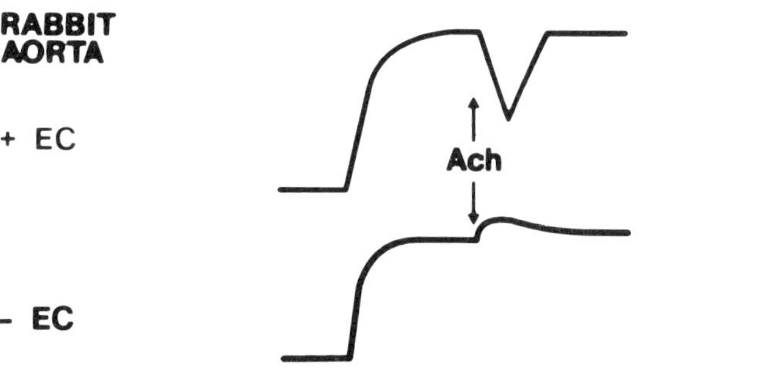

The importance of the endothelial cell for the vasorelaxant effect of acetylcholine (Ach). *When the endothelial cell layer is removed (− EC) acetylcholine produces only a slight contraction of an isolated band of rabbit aorta. In the presence of an intact endothelial cell layer (+ EC) acetylcholine produces a profound relaxation.*

Methyl arginine was used by Moncada et al. to provide one of the most important pieces of information so far about NO or EDRF. They showed that methylarginine in the whole anaesthetized animal causes a dramatic rise of blood pressure. This was long-lasting (Figure 19A), but could be reversed by an injection of the natural substrate L-arginine (Figure 19B). The same results that have been seen in animals have also been shown in man. This, then demonstrates the importance of NO in physiology. We are all living under a constant relaxation of our blood vessels due to a continuous outpouring of NO from the endothelial cells. A deficiency of this mechanism must surely lead to high blood pressure. Clearly, this exciting work will uncover new uses of nitro-vasodilators and new dilating drugs over the next 10–20 years. New anti-hypertensive substances will also emerge. However, this cannot be progressed by computer models or by single-cell preparations. Measurements of blood flow and blood pressure in whole animals will have to be made.

There is also much excitement being generated as the NO work expands, for we now know (Table 1) that NO acts as a second messenger, or cell-to-cell messenger, in many cell types. This augurs well for finding new drugs.

I hope that my varied examples of the way in which bioassay has led to important advances in medicine have been convincing. Not one of these discoveries, or many more, would have been possible without it. Nor would any of these advances have been possible without also working in whole animals to develop and extend the results found in isolated tissues. I have to emphasize most strongly that work on isolated tissues cannot in any way take over totally from work on the whole animal. For instance, blood pressure can easily be measured but the biology that contributes to its make-up is enormously complex and depends on the interactions of hormonal, nervous and other systems not available in a simple bioassay preparation. Thus, both *in vitro* and *in vivo* experiments are vitally important.

The last area of bioassay that I want to discuss is the clinical trial. No new drug can be marketed without extensive clinical trials to demonstrate its efficacy and safety and these are in essence yet another kind of bioassay where the activities of the drug are measured either against a placebo or

Figure 17

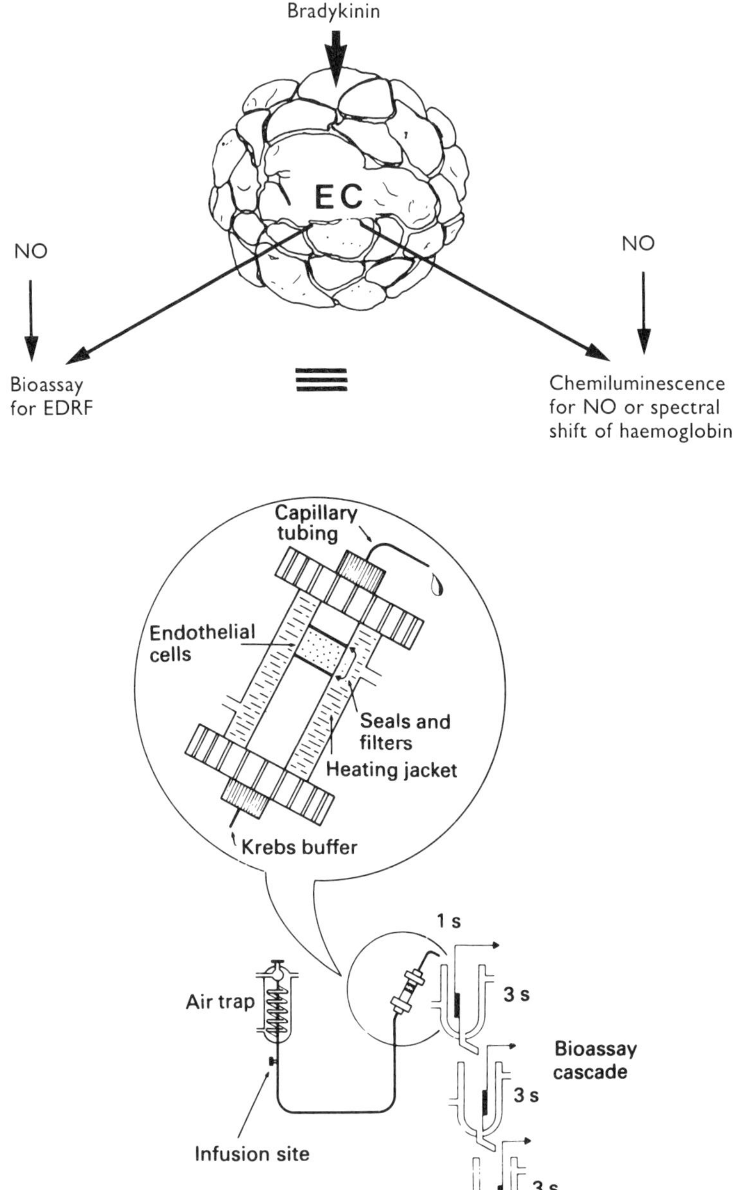

The generation of EDRF from endothelial cells (EC) grown on microscopic beads. *The cells were perfused and the perfusate passed over a bank of bands of rabbit aortic tissue (with the endothelial cells removed). The effect of EDRF, generated by addition of bradykinin to the cells, was compared with that of a solution of nitric oxide on both the assay tissues and by chemical measurement. From Gryglewski et al. (1986).*

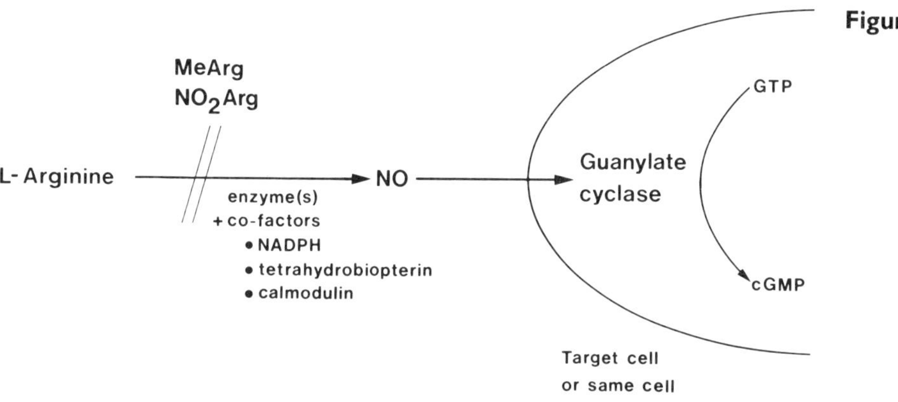

Figure 18

The formation and action of nitric oxide (EDRF) and its action on target cells. *Nitric oxide is formed from L-arginine and acts by stimulating guanylate cyclase with a consequent increase in intracellular cyclic GMP. The enzymic formation of NO from L-arginine is inhibited by methylarginine (MeArg) or nitro-arginine (NO₂Arg). Reproduced from Vane et al. (1990) with permission.*

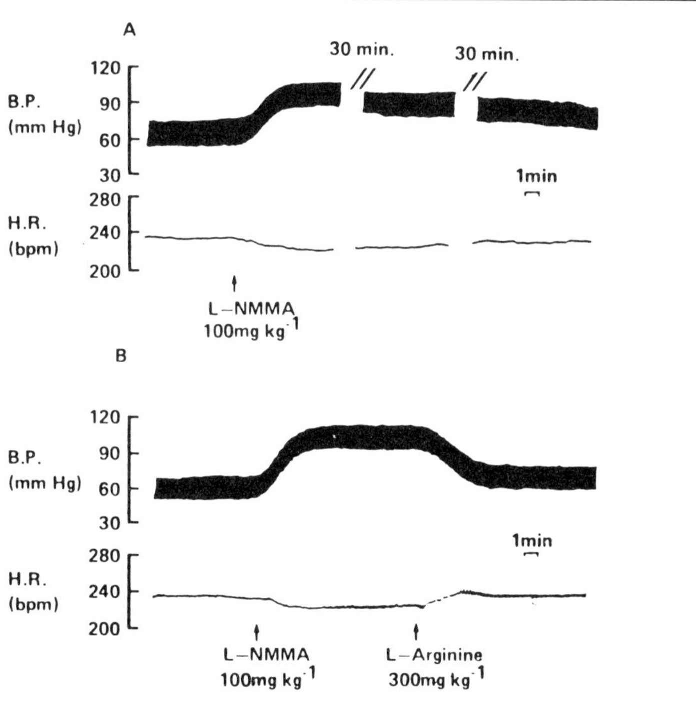

Figure 19

The effect of inhibition of nitric oxide formation by N^G-mono-methyl-L-arginine (L-NMMA) on the blood pressure of rabbit. *A single intravenous dose of L-NMMA resulted in a rise in blood pressure that lasted over 1 h (A). Administration of L-arginine (but not D-arginine) reversed the rise in pressure induced by L-NMMA (B). Abbreviations used: BP, blood pressure; HR, heart rate. From Rees et al. (1989).*

Table I

NO-producing cells	Effects of released NO
Endothelial cells	Smooth muscle relaxation
Macrophages	Cytotoxic (bacteria)
Neutrophils	Cytotoxic (bacteria)
Nerve cells	Nerve transmitter
Mast cells	Regulation of histamine release
Kidney epithelial cells	Second messenger
Hepatocytes	?
Kupffer cells	?
Astrocytes	?
Smooth muscle cells	?

Nitric oxide as an intercellular or intracellular messenger.

against another well-established drug. You will deduce from the principles which I enunciated at the beginning of this article that this type of bioassay demands extensive work in dozens or hundreds of people in order to make sure that the effects that are obtained are scientifically valid and statistically significant. Bioassay in man is complicated by elements which do not apply to bioassay in other situations. Just as measurements in the early days depended upon digits, hands and feet, measurements in the clinical trial area depend on attitudes and complex interactions between doctor and patient. In the 1950s, Beecher showed that clinical trials were confounded by the placebo effect which means that some 40% of people will react in the ways that they are expected to when presented with a new remedy. This placebo effect is due to diverse factors such as the enthusiasm of the doctor or nurse, bed rest, hospital atmosphere and so on. Pillars of the church are thought to be more susceptible to a placebo effect. What it means, of course, is that you cannot just give a new remedy to a group of people and expect to be able to tell whether the remedy works simply from the fact that you have up to a 40% improvement. Sugar would have done the same.

It is on this placebo effect that the enormous industry of fringe medicine is built. It is estimated that, in France, about a third of the medicines sold to the public have not been shown to be effective by properly controlled clinical trials; the same figure could well apply here. We are all susceptible to the fringe medicines, for we all believe that a coincidental effect (such as feeling better) could well be due to something we took that day. I myself am convinced, without a shred of evidence, that vitamin C in high doses will abort a cold.

Here, then, is an area of enormous concern which spills over into quack treatments for cancer and other unsolved diseases, because people turn to fringe medicines in desperation when conventional medicine fails to help. It is an area that needs control and we must all strive to make sure that the public is not duped by the escalating quackery which shelters under the overall umbrella of alternative medicine.

I hope that my examples have been enlightening and have shown that bioassay techniques in simple pieces of animal tissue have indeed contributed to medical progress. Without them, no new drugs and concepts of importance would have been developed. In just the same way, experiments on whole animals are also needed to develop and consolidate the initial findings and to make sure that the candidate drugs are as safe as can be predicted. Only after all of that, some 10 years later, can we turn to experiments in man which may well take another 5–10 years of intense effort.

One of the themes running through this article has been the ability of man to learn from his and others' experiments, and his and others' experiences. This learning process is formalized in science by publications in respected scientific journals. When concepts become established theories, they are also passed on to our children through the formal education process.

The animalists have, with some success, allied themselves to the environmental issue. It is unfortunate that the animalist argument is now presented to our children as a green issue, which it clearly is not. This inappropriate positioning is wrong and should be vigorously countered for there has been and will be an enormous world of discovery to which animal experiments contribute. Should the animalists succeed in their ambitions and stop animal experiments now, our children will be denied the medicines and cures which they have the right to expect in the next 10–20 years, arising from today's discoveries. But we cannot just leave it there. My own perspective of the animalist is that they should be positioned, not with the environmental issue, but with other zealots of extremist attitudes who are competing for the minds of our children. The conventional elements such as science, logic and civilized behaviour have to counter those sinister elements such as the destructive cults, quackery, animalists and terrorism. Those who advocate *only* the use of alternative methods belong with those who peddle alternative medicines and alternative religions, proselytizing as they do from the basis of belief, faith and mystique, rather than from the basis of tried and proven facts. We must do all in our power to ensure that young innocent and receptive minds are not filled with hocus-pocus, to be remembered as dogma in later years.

It is an unacceptable facet of modern day society that scientists who have worked all their lives to enhance the quality of life, to prevent suffering and distress and to combat disease should have to be advised to kneel at the end of the day, not to pray, but to inspect the under-sides of their cars for terrorist bombs.

We have gathered a wealth of experience from simple bioassay and from more complex animal experiments. The foundations for cures for cancer, acquired immunodeficiency syndrome, heart attacks, high blood pressure and many other crippling and killing diseases are already laid. We must ensure that these foundations are not wasted.

References

Bunting, S., Gryglewski, R., Moncada, S. and Vane, J. R. (1976) Arterial walls generate from prostaglandin endoperoxides a substance (prostaglandin x) which relaxes strips of mesenteric and coeliac arteries and inhibits platelet aggregation. Prostaglandins 12, 897–913

Burn, J. H. (1952) Practical Pharmacology, Blackwell Scientific Publications Ltd, Oxford

Gaddum, J. H. (1953) The technique of superfusion. Br. J. Pharmacol. 8, 321–326

Gryglewski, R., Moncada, S. and Palmer, R. (1986) Bioassay of prostacyclin and endothelium-derived relaxing factor (EDRF) from porcine aortic endothelial cells. Br. J. Pharmacol. 87, 685–694

Krieger, M., Salgado, H., Assan, C., Greene, L. J. and Ferreira, S. (1971) Potential screening test for detection of overactivity of renin-angiotensin system. Lancet i, 269–271

Rees, D., Palmer, R. and Moncada, S. (1989) Role of endothelium-derived nitric oxide in the regulation of blood pressure. Proc. Natl. Acad. Sci. U.S.A. 86, 3375–3378

Vane, J. R. (1957) A sensitive method for the assay of 5-hydroxytryptamine. Br. J. Pharmacol. 12, 344–349

Vane, J. R. (1971) Inhibition of prostaglandin synthesis as a mechanism of action for aspirin-like drugs. Nature (London) **231**, 232–235

Vane, J. R. (1972) in Pharmacology and the Future of Man. Proceedings of the 5th Int. Congr. Pharmacol. Karger AG, Basel

Vane, J. R., Angaard, E. E. and Botting, R. M. (1990) Regulatory functions of the vascular endothelium. New Eng. J. Med. **323**, 27–36

Yanagisawa, M., Kurihara, H., Kimura, S., Tomobe, Y., Kobayashi, M., Mitsui, Y., Yazaki, Y., Goto, K. and Masaki, T. (1988). A novel potent vasoconstrictor peptide produced by vascular endothelial cells. Nature (London) **332**, 411–415

7

Discussion

Michael Festing: I think it helps to put the numbers of animals used in medical research into perspective. This number represents only a very small fraction, probably less than 2%, of all the animals that we actually consume in a lifetime. I have calculated, for example, that the average person in this country will consume in a 75 year lifespan approximately 600 chickens, 20 sheep, 22 pigs and 4 cows. They will use in research four mice, one rat and one two-hundredth of a dog. In other words you can have all the benefits of medical research for a whole lifetime for approximately four mice and one rat. I think that's pretty good value for money.

Richard Jurd (University of Essex): May I make two observations. It appals me that writers in this field sometimes seem completely ignorant about basic facts of biology. For instance, I was asked by an acquaintance of mine who has written extensively on the subject of 'animal rights' whether insects had a nervous system! Also, in his writings, he groups insects, plants and stones together in what seems to be one taxonomic category and, on the basis of a particularly obscure point of biblical theology, separates mammals from birds, mammals being 'creatures of flesh and blood' and birds not being so.

These points would be ludicrous were it not for the fact that this writer is trying to tell those of us in biology and medicine how we should conduct our professional work and how we should be applying our studies to the causes that we feel are worthwhile.

My second point, speaking as one from a university which has a very radical tradition in this country, is that, talking to some of the 'animal rights' people within the university, it seems that the need to object is probably more important than the actual detail of the issue. They are simply objecting to something—never mind what. In a previous generation it might have been a totally different cause. I get the impression that certainly the leader of the 'animal rights' group in my university does not particularly like animals and that, perhaps 30 years ago, he might have been championing another cause. Perhaps the aim (although this might not apply to all) is not so much the cause of animals themselves but, through animals, a way of attacking what is seen as 'the Establishment', or powerful professions with vested interests, or Science with a capital 'S'—that is, anything seen as responsible for the world's ills. These two observations may help us a little in the assessment of the nature and motives of the opposition.

Maurice Miller (AMRIC): The Animal Rights Lobby of course takes advantage of the fact that people are generally sympathetic towards animal welfare—their dogs, cats, horses and so on; this cannot be disputed. The animalists therefore try to stand the truth on its head. They say that black is white; and the higher the profile they present, the greater the impact they make. Although they are not numerous there are enough of them to cause problems.

When I was a Member of Parliament, a woman came to one of my political surgeries to seek my opinion about animal welfare. We talked about the usual complaints. She said that she didn't use cosmetics and that she was a vegetarian. After a long discussion, I pointed out to her that 80% of animals used were rats and mice and she said: 'well they have souls too'. I asked: 'but what would you do if your house was overrun with mice?'. She responded with: 'Well, I have no problem with that; I keep a cat'. When the implication of what she had said dawned on her, she rose, strode to the door, turned and barked at me: 'I should have known to expect no understanding from a medical man!'

May I suggest that you keep AMRIC (Animals in Medicines Research Information Centre) in mind. We do a lot of good work, both on television and in schools, explaining the necessity of animal experimentation, taking into account the tremendous emotional overlay that is associated with the subject of animal welfare. *The Times* recently had a good article on this which suggested that, if you believe in animal rights, try telling the antelope that it has a right to live when it is being chased by a tiger or a lion.

What we can concentrate on, and I think this is the best course to take, is to get the GP involved. The General Practitioner in his surgery should be persuaded to put up posters, to have other notices about vaccination and so on, and to tell his patients every time a child comes in for vaccination, for some kind of innoculation or for medicine: 'you know this medicine I'm giving you will cure the child, or it will alleviate the ailment—and it has been tested on animals'—to tell them quite deliberately that without the use of animals their child might be dead.

8

Animal experimentation and neuroscience research

David Hubel
Professor of Neurobiology, University of Harvard

I address the problem of the animal activist movement here in the U.K. with some diffidence because I feel that, compared with the U.S.A., your experiences of it are more serious. I have often wondered why things are worse in Britain, and confess that I am still puzzled. There seems to be no doubt that it is true: objective comparisons, such as the terrorist activities here and in the U.S.A., would convince anyone that there is a major difference.

I became embroiled in these issues because in 1988/89 I was elected President of the Society for Neuroscience in the U.S.A. This is a society of some 16 000 neurobiologists. I decided at the beginning that the most important thing I could do in the Society was to take on the animal-activist problem as a kind of hobby, promoting a few ideas such as getting the message of the importance of animal research to ordinary people through their doctors, and getting the message to school children through their teachers. In the past couple of years I have had to slow down in this area, because I felt I should not give up science completely. That is of course exactly what the animal activist would like to see happen.

Neurobiology can be thought of as being subdivided into two main branches, one of which focuses on individual nerve cells, and the other—which we can refer to as 'systems'—which seeks to understand how these cells are wired together into circuits. The second branch depends on the first: we cannot hope to understand the circuits unless we first have a good idea about the elements from which they are formed. However, just as obviously, the cell-level work is not enough to allow us to understand the circuits, any more than the ceramics of bricks and mortar is enough to give us an understanding of the architecture of a bridge or building. The distinction between the two branches is important in considering the use of animals in research because much of the most important work at the level of the single nerve cell has been and continues to be carried out in lower animals which, by and large, only the most fanatical animal-rights people care about. A striking example is the research on the nerve impulse, for which Hodgkin and Huxley won the Nobel Prize in 1963, which was carried out on the squid. This creature happens to have an axon about a millimetre in diameter and could be investigated with techniques that were available at that time. The results have been shown to be completely valid all the way up the evolutionary tree. Work on systems can likewise be carried out in such relatively simple animals, with results that are of great biological interest. What we learn about circuits in a squid or grasshopper, however, does not necessarily apply to higher animals such as cats, monkeys or ourselves. If one's background is in medicine, as mine is, one is likely to be interested primarily in man and in medical applications; it may then be necessary to work on animals in which the circuits resemble those in man. I work on vision, and for my work the old-world monkey is ideal because as far as we know its vision is identical to ours. This holds for form, colour, depth and movement perception. Cats are suitable for studying some aspects of vision such as form and movement but not, for example, for work on colour, because the colour vision of cats is impoverished; it is not inaccurate to say that cats are not quite colour blind. Squids have vision that is entirely different from ours and what we learn by studying such vision, while fascinating, has little direct application to human vision.

Brain research in higher organisms can be justified in a manner convincing to any reasonable person. One has only to consider the spectacular progress that has been made in treatments for human nervous diseases in the past few decades. To mention just one class of brain diseases in which we have made progress, our knowledge of neurotransmitter function has led directly to the development of psychotherapeutic drugs which have changed the entire sociology of mental illness. In contrast to the days when I was a medical student, mental hospitals today are relatively empty.

The list of unsolved problems is even more impressive. Though we can treat their symptoms better than ever before, we still do not know the causes of most of the important psychoses, and cannot really cure them. We can do little about severe head injury, chiefly because we have so little understanding of why nerves in the central nervous system (as opposed to those in the limbs) do not repair themselves. We still do not know how brain circuits are formed, before birth or immediately after, or why the process of their formation often goes wrong, and of course such knowledge may be necessary if we are to be able to find remedies or preventions for congenital malformations of the nervous system.

When Torsten Wiesel and I began our research in the primary visual cortex of cats and monkeys in 1959 we were not thinking of practical applications. It was so obvious to us that the brain was worth studying that I suspect we did not even attempt to formulate explicit justifications. We were simply interested in how the brain worked, and the visual cortex seemed as good a place as any to begin. Yet it turned out that within a few years we were able to show that one of the commonest types of blindness, known as amblyopia, is a result of disruptions in the connections in the visual cortex. Our work not only helped establish the cause of amblyopia but also led to a means of preventing it. Until then most people had thought that the trouble was in the eye, and I doubt very much that we would have been funded had we said that we wanted to study amblyopia by recording from the visual cortex.

Ironically, studies of amblyopia, which are now being extended in many laboratories worldwide, are close to the top of the list of targets of the animal activists. The reason is that to bring about the condition one has to sew closed the lids of one eye. The procedure is entirely painless and does not seem to bother the animal at all. But when pictures of cats or monkeys with one eye sutured closed are prepared by the animal activist they can look disconcerting to someone who has no idea why the research is being done or how successful it has been. The activists certainly do not explain our motives, and instead imply that we do these studies out of pure sadism.

My reasons for carrying out brain research, I have to confess, go well beyond immediate medical gains, important as those are. If you ask what are the most interesting scientific problems that anyone could study I would answer that they fall into two large categories, the first concerned with understanding the universe and the second with understanding ourselves. Consider first the universe. About a year ago the Hubble space telescope was sent aloft to the tune of 2 billion dollars. My feeling about the venture was and still is one of total support because to me it is one of the most exciting things human beings can do. It is an example of how far we have evolved intellectually, that we can take an interest in such things as quasars. Yet you would have to stretch your imagination to suggest any very practical applications from such work. The major motivation for it has been scientific curiosity, the desire to know about the universe around us—exactly the same urge that motivated Newton and Galileo. No doubt other branches of science will gain from such studies, and inventions that arise as a by-product of the technical advances will benefit humans, but these considerations are seldom uppermost in the scientists' minds. Similar things can be said for our huge accelerators, for experiments in high energy physics, and no doubt for many other branches of science.

Curiosity also motivates our desire to understand ourselves, and undoubtedly one of the most fascinating aspects of man is his brain. It is the most complex known structure. The things it does can be incredibly complex, even if sometimes incredibly foolish. How could studying it not be one of the most interesting of all possible occupations, especially today, when we find ourselves suddenly with the technical equipment to make such a study feasible?

Yet when we go about seeking the resources to study the brain we tend to express ourselves entirely in terms of practical applications and, in particular, practical medical applications. Astronomers have no such recourse and must justify their activities mainly on grounds of scientific curiosity. To judge from their budgets, they seem to do this very well, I would be the last to deny the importance of medical applications, but we short-change ourselves by always couching our grant requests exclusively in those terms.

We limit ourselves in yet another way if we justify our work only in terms of immediate and direct medical gains, because the possible benefits will go well beyond medicine. Think of the advance we would gain in a field like education if we had a detailed knowledge of brain mechanisms of learning, or in combating criminal behaviour if we understood the brain's role in emotions and rationality.

The ethical problems of carrying out medical research involving animals have been mentioned earlier and in general I think the ideas that have been expressed are close to mine.

The arguments of the animal activists are divided into two categories: the profound and difficult on one hand and the fatuous on the other.

First, the more difficult and older argument is that we as humans have no right to assert our domain over animals. To my mind, the very words in which this argument is couched portray a complete and profound ignorance about the way animals exist and compete on our earth. One may or may not like it, but it is certainly clear that many animals are either predators or prey, or both. Many of the animals in this world are obligatory carnivores. If you own a cat and want to see it thrive you don't get very far trying to feed it carrots; you have to feed it meat. A few years ago my son owned a boa constrictor, an impressive and beautiful animal, which luckily did not grow too large. To stay alive that boa had to consume a mouse every month. To watch the boa go after the mouse was, in some macabre way, rather an aesthetic experience. You may not like the idea of a boa eating a mouse, but if you feed the boa a piece of meat or a dead mouse or anything apart from a live animal, you will end up watching it starve.

Quite aside from predation, we know that if we do not do something to get rid of many of the rats and cats that roam our streets we will be over-run by them. Thus, it is common practice in any advanced nation to take a certain proportion of our stray cats to the pound and have them humanely killed. If that is not done our cities will be a mess and the cats will starve. Surely it is more humane, if that is the right word, to kill them with a painless barbiturate injection than to let events take their course without intervention.

Observing a phenomenon like the boa eating the mouse is an educational experience from which most people in our society are today excluded. Our children perhaps should see the real world as it is rather than remain insulated in what is today something of a glass house. Over the last century our society has undergone quite profound changes in this respect. When I was a young boy in Montreal, horses walked up and down the street pulling wagons and sleds, urinating and defaecating in the process. That was part of life. If your grandmother died her remains were kept at home until the funeral when she was finally taken off to the cemetery. Now children aren't exposed to that sort of thing at all. A relative that dies is taken straight away to what in the U.S.A. we call a funeral home; neither the relative nor anyone else wants to know exactly what happens there. One is completely insulated from these real life experiences. No wonder that all of us, including the animal activists, are disconcerted at the idea of a surgical procedure, whether it is performed on a human or on a research animal. It doesn't matter that the human or animal is under an anaesthetic from the beginning. To the experimental animal, for example a cat, it makes absolutely no difference whether it is given the anaesthetic at the animal shelter (which is the U.S. euphemism for a pound) and immediately killed, or given an anaesthetic in a laboratory and killed after an experiment is done. But because of the efforts of the animal activists it is impossible in England, and in many states in the U.S.A., to obtain animals for research that are destined to be killed in pounds or dogs' homes. The result is that a cardiac surgeon who wants to practise his operations or to develop new operative techniques on dogs must now purchase a purpose-bred animal for $600 rather than the $30 that was the cost when I was a medical student.

Thus, the animal activists' cause is fuelled in essence by a failure to understand, in real terms, how the animal kingdom works and an excessive squeamishness that is the product of the insulation produced by our advanced technological civilization.

So much for the more difficult ethical considerations. Now we have silly or simply false arguments. We have the argument that medical research is on the whole unsuccessful, that no major disease has ever been cured or alleviated by medical research.

Any schoolboy can satisfy himself of the falsity of this assertion by spending a few hours in a library, but the activists are interested in developing a following, not in the truth, and evidently not in saving human lives.

A second argument is that all medical research can be carried out at the level of the cell or tissue in culture. People must be made to understand that although such techniques are very powerful for addressing problems at the cell or tissue level, they cannot be used to address problems at the organizational level, just as you cannot understand a bridge by understanding the ceramics of bricks. Of course the other argument is that you have to get the cells from somewhere, namely from animals such as mice.

The third argument is that the problems in biomedical research can be resolved with computers. I would be the last to deny the usefulness of computers because I use them in my laboratory and in fact spend an inordinate amount of my time programming them. But I use them as tools, not as a substitute for the brain. The fact is that, at present, computer engineers tend to use the information they get from the neurophysiologist in constructing their computers much more than the other way round. Nevertheless, I do think that one of the most interesting developments in the last 10 years has been the use of computer models to imitate the way that the brain might work. This is particularly fascinating to me in the realm of learning because machines can be developed that, in some sense, learn by themselves. I think such work holds much promise for the future, but now it is still in its infancy.

Similar arguments apply to areas of medical research other than brain research. I have just come back from Boston, where I had a coronary by-pass operation, and I must say I am very pleased that my surgeon didn't develop this operation on people like me, or practice it on a computer. The importance of animals for developing and for practicing new methods in surgery seems so obvious, and yet the people who carried out the research and those who apply it every day sometimes seem to be in a conspiracy to keep it a secret. The other day in a chemist's shop I picked up several pamphlets, put out by the British Heart Association, and in an otherwise excellent account of heart research I searched in vain for any mention of the fact that animals are absolutely essential for the enormous progress that is being made. It is as if we wanted to hide this fact from people. How can we be surprised when people fall for the specious propaganda of the animal rightists?

Since these arguments of the animal activists are, as I have said, manifestly silly, one wonders why they are sometimes believed (and we have to assume that they are believed since they continue to be used). One reason for this is that the standards of scientific education, at least in the U.S.A., are low. As a vivid example of this one only has to cite the well-known fact that the wife of a recent American president arranged his timetable by consulting an astrologer!

The discussions for and against the use of animals in research often take the form of a formal debate. My feeling is that a debate is basically a game, the purpose of which is not to establish the truth but to prove whether or not you are a good debater. Thus, during a debate, one side says one thing and the other side says the opposite and the listener (perhaps a member of the press) tries to arrive at the truth by a process of adding up the two arguments and, as it were, dividing by two. The only debate that I have made the mistake of engaging in was one in which my opponent was the president of an American organization called 'Physicians for Responsible Medicine'. This group consists of several thousand physicians, which sounds like a large number but is a minuscule proportion of the perhaps half a million physicians in the U.S.A. My discussion with this gentleman took the form of his saying something, and my immediate response: 'That's not true'. The same assertions and denials were repeated throughout. One of his favourite statements was that animals used in research suffered terribly. I pointed out that 99% of them don't suffer at all: we have to get approval from federal, state and local committees to do the work, and any suffering is allowed only if it is absolutely necessary to the experiment—and then we must use sedatives to alleviate any distress. Nevertheless, my adversary simply reiterated that the animals suffered terribly—and I kept denying it. What is the poor person external to the argument to believe?

A further example was my pointing out that during the Nazi era one of the major forces in Germany was the animal rights movement, stemming doubtless from the fact that Hitler loved pets and that the Nazis clearly thought much more highly of their animals then they did of their fellow humans. The immediate response to this statement was: 'That's not true'. Of course I had heard these facts about Nazi Germany but hadn't examined the evidence myself and could not cite chapter and verse. Afterwards I went to the library, examined the literature on the subject, and thoroughly satisfied myself as to the Nazi's attitude to animals and, in particular, the use of animals in research. It is very convincing to read the newspapers that came out at the time, especially the published transcriptions of speeches of the high officials of the Nazi regime.

(I am sorry that at the time of that debate I didn't know about Heinrich Himmler's dog. As you may remember, prior to World War II, Himmler was the Third Reich's ambassador to England, and the embassy was the building that now houses the Royal Society of London. Outside the building, in a tiny plot of grass, one can still find a small tombstone marking the grave of 'Ciro', Heinrich Himmler's dog. My point is not that there is anything wrong with loving animals, or even according them full-blown funerals, but only that Nazis, like many animal rights activists, seemed to have loved their animals more than their fellow humans.)

To conclude, I would like to consider what we may usefully do to combat the problems posed by the animal rights lobby. When I was President of the Society for Neurosciences my main compaign was to get physicians interested in this question. It seemed to me that if each doctor would say just a few words to each patient, if there were perhaps one relevant sentence at the bottom of the prescription pad, if hospitals and doctors' waiting rooms had on their tables pamphlets such as those produced by the RDS here in Britain and the Foundation for Biomedical Research in the U.S.A., if relevant posters were put up on the walls of hospitals' or doctors' waiting rooms, the impact would be enormous. Doctors have the almost unique advantage that they can get to huge numbers of people. Unfortunately I was disappointed with the results of my campaign. I wrote letters to every member of the Society. I made speeches and got very positive results in terms of feedback at the time, but my feeling in the end was one of discouragement.

For example, I wrote to the directors of every hospital in Boston with the suggestion of posters and pamphlets in waiting rooms and corridors. I got an answer from only one hospital director. The one man who answered was very positive and very encouraging. This was 2 years ago. The positive response was, as it happens, from the hospital that I went to for my cardiac by-pass. I spent 2 delightful (more or less) weeks there, but on no wall did I ever see a poster on the use of animals, and on no table did I see any literature. Perhaps a large part of this is simply passiveness—doctors and administrators have busy schedules and many things on their agenda, and thus never get round to this particular action. But there is also a strong reluctance to stir things up. The medical community has yet to realize that keeping a low profile as a strategy has not worked. For my part, I realize now that what I should have done was to send out all of the letters, and then phone each hospital director and talk to him, or go and visit him. I should have got to the people engaged in research in those hospitals, urging them to exert pressure on doctors and administrators, who could possibly be made to feel guilty at using every day the result of the research that they are unwilling to defend. I also made an effort to encourage scientists to help educate young people. It is of crucial importance to get into the schools. This is something that our opponents are doing with enormous success. Meanwhile we are sitting by and not even presenting our side of the argument. And our side is so compelling that even a child will come to the right decision. We need to try to educate the editorial boards of newspapers, television station managers and all of the other people that are in direct contact with the public. I made formal approaches to two illustrious American newspapers. In one case I was delighted with the response; in the other I was dismayed since in that case the editor of the business page was frankly antagonistic to our cause. You cannot expect that you will always have a sympathetic audience, but at least you can state your case.

Another area in which we fail, in the U.S.A. in any case, is in our organization. Our Foundation for Biomedical Research has a capital that amounts to something like a few hundred thousand dollars. Against that we have some 500 organizations that are working on behalf of animals, the most important and strident of which is probably People for the Ethical Treatment of Animals (PETA) whose capital amounts to many millions of dollars. If our potential supporters would rally around our Foundation for Biomedical Research in the U.S.A., our strength would be overwhelming to almost any set of opponents. By 'potential supporters' I mean the pharmaceutical houses, the hospitals, medical schools, the professional organizations (in medicine, surgery, pediatrics and so on), the academies (the National Academy of Science, for example), health charities and, not least, the patients, particularly the incurably ill patients. Taken together, the power of all of these would be enormous. We will need that power. The animal rights organizations in the U.S.A. (and everyone knows that the U.S.A. is a very lawsuit-conscious country) are catching on to the importance and the power of lawsuits. A person wrote a very reasoned editorial in a Washington newspaper a year or so ago dealing with what are known as the Silver Spring monkeys. She was immediately sued by PETA. Animal rights organizations do not seem to care much whether they win or lose. Their ambition is to put their opponents out of business, and their frequent success is making us much more reluctant to say things in a forthright way.

We have to be much tougher and more forthright in our dealings with the animal rights people. I think we have to be bolder and more emphatic, and the higher up you are in the hierarchy of organized medicine, the more incumbent it is to be forthright and bold. For physicians not to support animal experimentation whilst they are using treatment derived through such medical research is simply immoral. And they should be told so in those words.

9

Animal experimentation and cancer research

Sir Walter Bodmer FRS
Director General, Imperial Cancer Research Fund

Introduction

The first Director of the Imperial Cancer Research Fund, Ernest Bashford, was appointed to his post at the turn of the century largely on the basis of a draft scheme he produced on 'Enquiring into the Nature, Causes, Prevention and Treatment of Cancer'. The scheme included many suggestions that would still seem appropriate to cancer researchers today: there were recommendations for extensive statistical investigations involving the study of human populations, for example, as well as proposals for experiments on animals. In fact, Bashford and his colleagues were pioneers in the study of the transplantation of tumours even though they lacked the wide range of techniques available to today's researchers and so had only limited experimental approaches open to them. Bashford's investigations were bedevilled by the complete lack of knowledge then about the problems of incompatibility and transplant rejection.

Nevertheless, it was the sorting out of these problems that eventually led to our modern understanding of how the immune system distinguishes self from non-self and to the development of inbred strains of mice such as are used now in biological research.

The aim of this paper is to describe some important lines of investigation in cancer research and to emphasize the enormous contribution that animal experiments have made—and, I believe, will continue to make—to our present understanding of the nature of cancer. I do not intend to say much about the treatment of cancer even though each of the main approaches—surgery, radiotherapy and chemotherapy—has depended in one way or another on animal experiments. In the case of chemotherapy, for instance, it was Skipper's classic work in mice that provided the basis for the chemotherapeutic protocols that have successfully cured more than 50% of all leukaemias in children.

Prevention and treatment

One of the numerous criticisms levelled by antivivisectionists (or 'animalists') against the use of animal experiments in cancer research is that the experiments are, in any case, only directed at treatment rather than, what must surely be the primary aim, prevention. Although this is true of many uses of animals in cancer research, there are important lines of research—for example, on promotion, vaccination and immune response—that do depend on the use of animals. Very often those that make this criticism of animal experiments in cancer research consider that what is being done is merely 'stuffing chemicals into mice and seeing whether they get cancers'. This cannot be further from the truth, as I shall explain.

Prevention of cancer is obviously important and, in principle, should involve no more than identifying the factors involved and then controlling them. However, although this approach sounds logical and simple, the difficulties involved should not be underestimated. The correlation between cigarette smoking and lung cancer, for example, shown through the work of the late Bradford Hill and Richard Doll, has been established for nearly 40 years. Yet, even when a clear-cut environmental factor has been identified as a cause of cancer, persuading people to use the knowledge can be a formidable task. The link between cigarette smoking and lung cancer has been publicized extensively, but smoking is still a problem in our society: preventability and prevention are two very different things.

Our knowledge about the damaging effects of cigarette smoking has been aided by the classic experiments carried out by Berenblum on mice. He showed that there was a distinction between initiation (starting the process that actually triggers the formation of a cancer, usually with some sort of mutagenic agent) and promotion (maintaining its progression, by agents which have some effect on the growth of cells). We know now that cigarette smoke contains both initiators and promoters. This is why, when you stop smoking, you stop any increase in risks subsequently and why, as far as lung and other smoking-related cancers are concerned, there is considerable benefit in stopping smoking.

Figure 1, taken from the well-known survey by Doll and Peto, illustrates the importance of various environmental factors as causes of cancer. The Figure shows that, after tobacco, diet is probably the most important factor. However, despite a variety of studies, we are still to identify the dietary factors that most influence the incidence of different cancers and so cannot yet be sure about which components of the diet are most significant. There is no doubt that it will be through further fundamental understanding of cancer as a disease, and by setting up and studying appropriate models, that we will ultimately learn which specific components of the diet are responsible. Figure 1 also shows that viruses (which are probably now responsible for 10–15% of cancers worldwide), particularly the papilloma wartlike and the hepatitis viruses, are an important cause of cancer. The former cause cervical cancer, which is very common in parts of the world outside Europe and the U.S.A. Our understanding of viruses as an environmental cause of cancer began with fundamental studies in animals. Ultimately, the problem of virus-induced cancers will be combated by applying knowledge derived from animal experiments and modern techniques of molecular biology together with an improved understanding of the immune system.

Viruses and oncogenes

To be able to prevent and treat cancer, it is crucial to understand its fundamental causes. The basic ideas about the causes of cancer are not particularly new. They date back to the great surgeon, Percival Pott, whose observations in the 18th Century on chimney sweeps suggested an association between cancer of the scrotum and soot. This association was eventually established in the early years of this century by the demonstration that there were certain chemicals in tar products which, when given to animals, produced cancer. Over the years we have been able to identify a growing number of cancer-

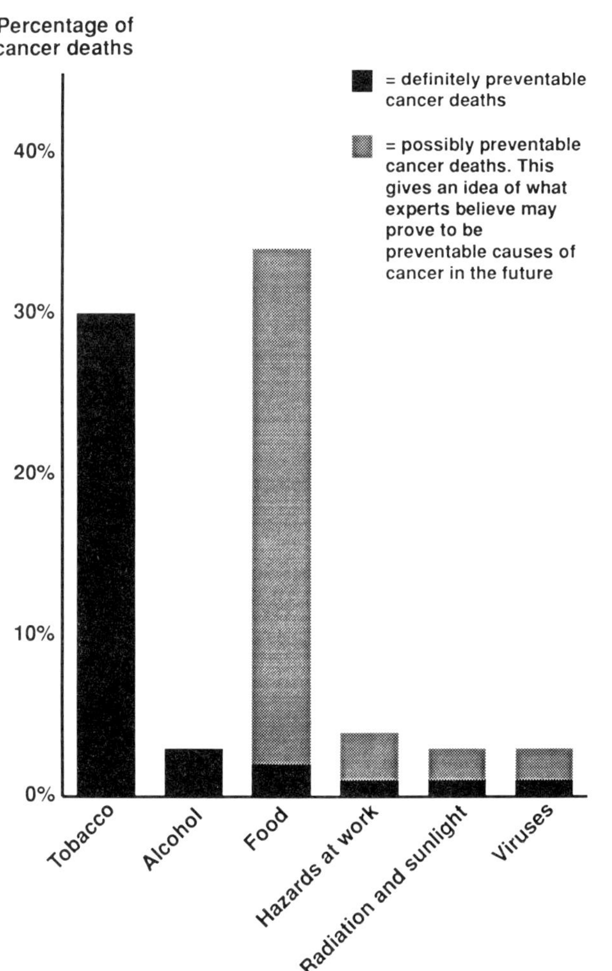

Environmental factors as causes of cancer (Doll and Peto)

causing chemicals (carcinogens). We have also begun to understand the role of the immune system in the development of cancer, an understanding that began with simple experiments into the way animals respond to infection.

Our understanding of cancer as a series of genetic changes that gradually take cells from the normal to the abnormal has its basis in the work of Peyton Rous. He was the first to show that a fowl sarcoma could be propagated by means of a cell-free system; i.e. the sarcoma was caused by a virus. Peyton Rous' work, though ignored for many years, attracted much attention in the 1950s when there was a great deal of interest in viruses. However, people became disappointed when it was apparent that most human cancers appear not to be caused by viruses, at least not by the sort of viruses that Peyton Rous had studied. Nevertheless, his research led to our most fundamental understandings of the genetic steps involved in the development of cancer and also to totally new experimental approaches that we can now envisage using in both prevention and treatment. Moreover, it was research work on viruses, using animals in the first instance, which led to the discovery of the acquired immunodeficiency syndrome (AIDS) virus.

Before discussing particular areas of cancer research, it is relevant to emphasize again the importance of the fundamental work that underlies it, and the contribution of animal experimentation that is rarely noted. It is no exaggeration to say that the whole development of molecular biology, leading to the remarkable knowledge we now have about the structure and function of DNA, has depended on animal experiments. One of the most famous of these was the experiment of Griffiths in 1928. Griffiths showed that, if a strain of pneumococcus which was lethal to mice was killed by heating, it could still cause death if it was given to mice at the same time as a non-lethal but living strain. This illustrated that something was being transferred from the heat-killed, virulent strain to the non-virulent strain. Griffith's experiments, totally dependent on the use of animals, led to one of the most fundamental contributions to our understanding of how DNA works.

Working with viruses necessitates the use of tissue culture techniques—yet another advance which has depended on animal experiments. The development and refinement of tissue culture techniques has depended on knowing about vitamins and antibiotics, knowledge obtained from animal experiments. Tissue culture is widely quoted as one of the alternatives for animal research. It is not an alternative, of course. It is simply one of the most fundamental tools that is used in any cancer research and indeed in most other areas of biological research today.

Viruses can cause cancer-like transformations in tissue culture. Figure 2 shows the very distinct morphology of a colony of cancer cells. Such a transformed colony, when put into an appropriate animal host, will cause a cancer. Having such a system, that can be conveniently studied in a laboratory, makes it possible to determine what particular feature of the virus is responsible for the transformation, i.e. what particular feature causes cancer. The ability to recognize that feature has resulted from applying recombinant DNA techniques to the analysis of the class of viruses that Rous first showed could cause cancers—those viruses that have RNA as their genetic material.

The essential gene content of these viruses can be analysed. The canonical virus, for example, has three main genes: one for an internal set of proteins, a second for the coat protein and a third for an enzyme that replicates its nucleic acid. The cancer-causing viruses have extra genetic information (the oncogenes) that is essential for their ability to transform normal into cancerous cells. This extra information is an altered form of genetic information found in normal cells. The clue to its function came from studies on the control of cell growth, in particular on the growth factors that trigger cells to divide.

The first of these factors, epidermal growth factor, was first detected by the fact that it caused premature opening of eyelids in new-born mice. Growth factors are proteins that, when attached to a particular receptor on the surface of a cell, trigger that cell to divide. As the function of growth factors became known, it was a natural thought that if something were to go wrong with this triggering process—for example, if its production were not controlled effectively—then this could be a key step in the development of a cancer.

Work therefore began on the study of the primary structure of growth factors and their receptors. Mike Waterfield, at the Imperial Cancer Research Fund, first showed that some oncogenes (i.e. altered or mutated

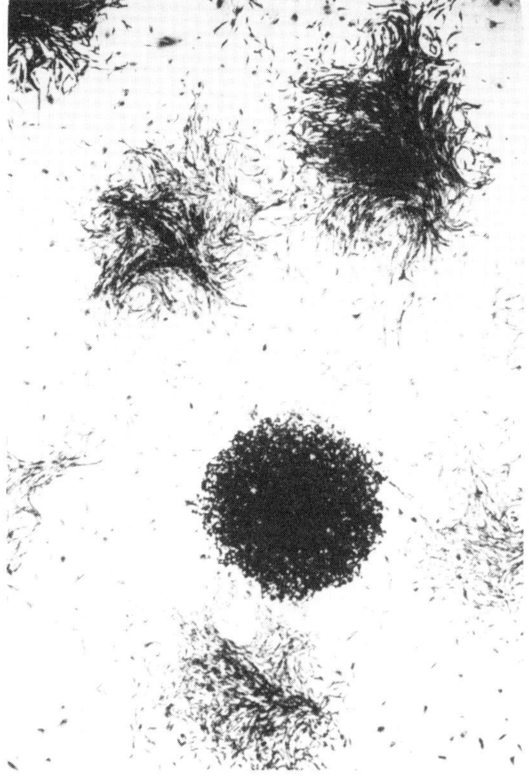

Figure 2

Illustration of the morphology of a colony of cancer cells.

forms of genes that occur in normal cells) seemed to produce altered versions of either growth factors or their receptors. This makes sense since there are simple ways in which alterations in growth factor function could encourage the growth of a cell. Normally the process of growth is balanced naturally with growth factors for one type of cell being produced by another cell type. However, when a cell can produce its own growth factor, in a kind of narcissistic way, then that cell is able to divide without external controls, and that is a step on the way to cancer.

Thus, abnormal production of a growth factor by the cell itself (an 'autocrine process') is one of the key steps that can take place in the development of a cancer. Alterations in a receptor so that it is continually activated can be another mechanism by which cell growth is maintained in the absence of normal controls.

As soon as mechanisms such as these are elucidated, it is then possible to look for ways of blocking function in the manner that has been described so elegantly by Sir John Vane. A new approach to the treatment of some cancers, therefore, involves using classic pharmacological procedures for blocking the functions of growth factors. An example, again arising from work at the ICRF, involves a 'bombesin-like' peptide which has been shown to be a growth factor for a form of lung cancer. Bombesin itself is a molecule which was first identified from work on frog skin, but it has an analogue, gastrin-releasing peptide (GRP), that occurs in man and other mammals. GRP is an autocrine growth factor for certain lung tumours and it has been suggested that blocking its function may be an effective way of treating the approximately 25% of lung cancers in which this mechanism is involved. Investigations into ways of blocking its function can start with cells in culture, but they must proceed through animal experiments to human trials before any successful treatment using this type of approach will be seen.

As well as providing a basis for the treatment of certain types of cancers, it is possible that the release of growth factors during the early stages of tumour development might also provide a basis for early detection. Early detection of cancer is very important because we know that the earlier it can be diagnosed the more likely it is that a cure can be effected.

Cancer genetics

I would now like to move on to consider ways of finding out more about the genetics of cancers. One way to do this is to study cancer in human families. Although cancer is largely an environmentally caused disease, there are well-defined, familial versions of some common cancers, for example bowel cancer. Bowel cancer is the commonest cancer not attributable to smoking in many Western countries, killing about 20000 people each year in the U.K. (Improvements in treatment have been much less than one might have hoped over the past 20 or 30 years.) About 0.5% of all bowel cancers involve an inherited susceptibility, called adenomatous polyposis coli (APC), in which affected individuals have up to 1000 or more pre-cancerous growths or polyps in their bowels. These polyps each have a small chance of developing into a cancer so that the overall probability is that an affected individual will be certain to develop one, if not several, cancers of the bowel unless the risk is removed by removal of all or part of the bowel.

Nowadays the best approach to investigating such an inherited disease is to try to locate the gene. Figure 3 illustrates that the gene responsible for APC has been localized to a small region of chromosome 5; in this region, in rare cases, patients with polyposis have a small part of this chromosome deleted. Knowing this, and using the most modern techniques of genome analysis, it becomes possible to try to isolate the gene itself and to determine its function—knowledge which may provide a variety of clues as to how the activity of the gene can be prevented in people with the inherited susceptibility to APC. It will also be possible, using transgenic mice, to create precisely the same mutant in the mouse and so have an appropriate animal model to study. This will not only be useful for studying approaches to treatment but also for studying possible ways of preventing the disease. For such a common cancer this is of crucial importance.

Having localized the APC gene to chromosome 5, it has also been possible to show that the same gene is involved in more than 60% of all bowel cancers, most of which are not connected with any major inherited susceptibility. So, finding the precise defect of the gene that causes the rare, familial, inherited disorder will also provide an understanding of the aetiology of the majority of bowel cancers, among the commonest of all cancers. Progress along this path has been made by the use of sophisticated molecular techniques in combination with animal experimentation.

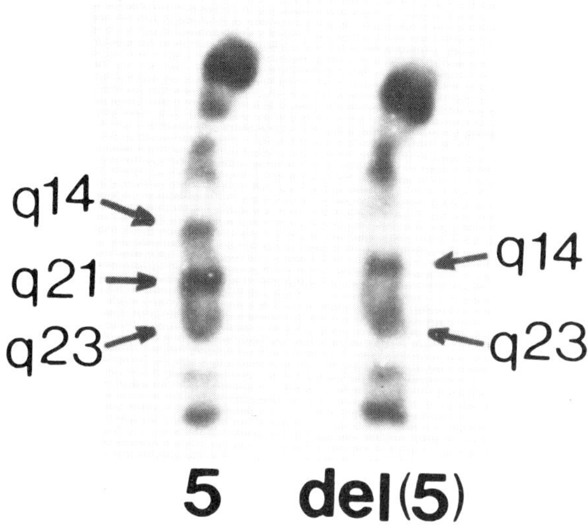

Localization of the gene responsible for APC to a small region of chromosome 5.

Genetic approaches such as these have led to the uncovering of a different class of genetic changes from those brought about by oncogenes. Oncogenes have a 'dominant' effect because once the gene is there in an altered form it causes a step in the development of a cancer even when the normal version of the gene is still present. The other class of genetic changes can be considered 'recessive' because the gene mutation results in the absence of a factor that normally prevents the cancer from developing. Thus, the function of both genes normally present in a cell needs to be knocked out.

One such gene, called p53, was discovered through work both at the ICRF and in the U.S.A. It is now known to be the gene that is more frequently mutated and changed in all human cancers. The p53 protein was found because it associates with a protein produced by one of the viruses that was studied classically as an oncogenic virus. It is significant that this is one of a class of viruses whose genetic material is DNA—like the wart-causing papilloma viruses involved in cervical cancer—rather than RNA, as in Rous' original virus or human immunodeficiency virus. Generally speaking, apart from papilloma viruses and the virus that causes mononucleosis, DNA viruses have not been implicated to any great extent in causing human cancers. But again, through studying these viruses, we are learning a great deal about fundamental changes that take place in cancer cells.

Other classes of cancer-causing viruses also involve the same p53 protein, as well as another protein called RB. RB stands for retinoblastoma, a tumour of the eye, the susceptibility to which is also inherited. It was through genetic studies that this protein was first identified and then shown to bind to the transforming proteins of certain viruses which cause cancer in animals. Once again, it is the animal work leading to the understanding of viruses and the way they cause cancer that led to these discoveries.

Immune response

Studying the changes produced by mutations in genes is important for another reason. They may provide targets which the immune system of the body uses to attack a cancer. The notion that the immune system, which protects us from infections, may also be important in controlling the growth of tumours, is an old one which became discredited when it was found that suppressing the immune system of an animal or a human does not seem to lead to any major increase in the incidence of cancer, apart from certain cancers caused by viruses. It was also thought unlikely that the sort of changes known to occur in cancers would be 'seen' by the immune system, since they were thought not to be manifested on the surface of cells. We now know that this is not so, because of the dramatic advances in our understanding of immunology.

It was skin-grafting experiments in animals, pioneered particularly by Peter Medawar, that led to much of our understanding of how the cellular part of the immune system works; namely, how cells, rather than antibodies, attack foreignness. This in turn led to our understanding of the major histocompatibility system—the human leucocyte antigen (HLA) system—a collection of molecules on the surfaces of cells which have to be matched for any successful transplantation. These molecules play a key role in the interactions between lymphocytes and the cellular targets they attack by somehow presenting antigen to the lymphocyte receptor that selectively recognizes foreignness. Our understanding of this process has been revolutionized within the past few years, particularly through the work of Alain Townsend and his colleagues. Studying the response of the mouse, and then of man, to the 'flu' virus, it has been shown—quite remarkably—that even proteins within a cell, when they are broken down and re-presented on the surface, can be recognized by this system. What this means is that viruses cannot hide within a cell. Their protein components form a sort of molecular 'signature' that can get left on the cell surface and be recognized by the immune system. The understanding of this process is leading to a completely new approach to the production of vaccines. More importantly, from the point of view of cancer, it has led us to realize that some of the changes within a cell may also be recognized by the immune system and so could be targets for therapy.

There is some evidence for this in that many tumours have been shown not to express on their surface the HLA molecules that are required for the process of T-cell recognition. Figure 4, for example, illustrates the use of a monoclonal antibody to stain normal tissue that expresses a particular antigen. Tumour tissue does not stain; it has escaped immune attack by the body since, because of a mutation, it does not express the molecules that the cells of the immune system can recognize. I have no doubt that, in time and as a result of work with mice that have had parts of the human system implanted, we will be able to devise procedures for attacking cancers that use this knowledge of the immune system.

Monoclonal antibodies

The immune system is useful as a tool in many ways, particularly through the production of monoclonal antibodies. This involves fusing antibody-produc-

Figure 4

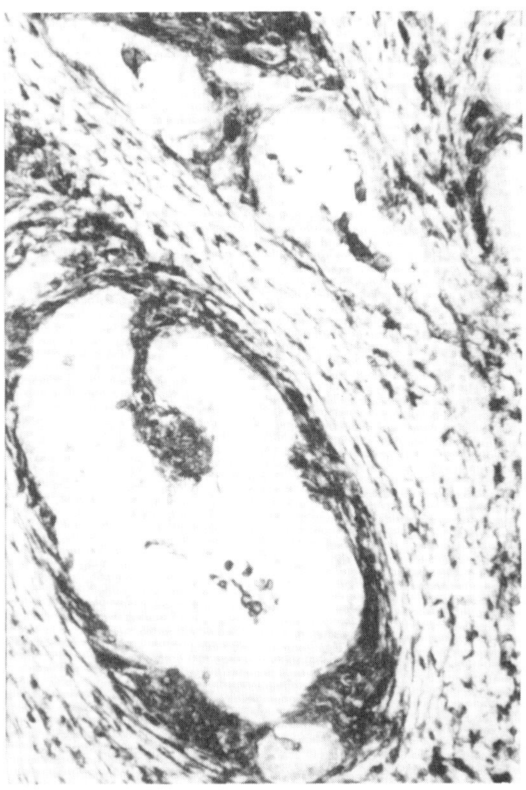

The use of a monoclonal antibody to stain normal tissue that expresses a particular antigen. *The unstained area is tumour tissue.*

ing cells from the spleen of an animal that has had, for example, human cancer cells administered, and obtaining, in tissue culture, a colony of cells that produces a single type of antibody molecule. For example, this may be an antibody specific for one particular chemical feature of a cancer cell. It may soon become possible to by-pass the use of animals for achieving this but, if the techniques for monoclonal antibody production had not been developed in the way they were, using mice and tissue culture, it would not now be possible to attempt to take the further step of producing monoclonal antibodies without using animals.

Monoclonal antibodies have found many uses in cancer research, from diagnosis in the pathologists' laboratory, to the detection *in vivo* of metastatic growth and approaches to therapy which involve targeting toxins and radioactivity to tumours. The development of monoclonal antibody techniques in cancer research depends on the ability to grow human tumour cells in mice, a possibility that has arisen because of the existence of a strain of mice, the nude strain, which is almost completely defective in its immune response. Since these animals lack an effective immune system, a human tumour can grow in them without rejection. The current widely used technology is to grow human tumours in these nude mice and to use them as an experimental vehicle for trying out novel approaches to diagnosis and treatment. If you take a monoclonal antibody that will bind to a molecule on a human tumour cell implanted into a mouse, and label it with a radioisotope, then you can monitor it, with appropriate imaging equipment, to find out where the antibody has accumulated. Such experiments were first carried out in mice about 10 years ago, and the knowledge obtained by developing and refining the imaging technology using these animals has now been applied to tumours in humans. There are now some situations where such imaging of tumours in patients is an integral part of the diagnostic process.

If you can get good images of a tumour using a monoclonal antibody, then you should also be able to deliver treatments to selected target areas. These treatments could, for example, be in the form of radioactivity, giving a kind of targeted radiotherapy. Such an approach is already being used, with initial indications of success, to treat some brain tumours. In one case, for instance, the antibodies are injected into the cerebral spinal fluid so that they have direct access to tumours in the brain.

Another approach is to attach cytotoxic substances to an appropriate antibody molecule. For example, you can attach ricin, which is a powerful plant toxin, to an antibody and, provided it is securely attached and the antibody is selective for cancer cells (and not for too many normal cells) then it is possible to deliver the toxin, via the antibody, directly to the aberrant cells. Experimental mouse systems have shown this approach can work remarkably efficiently. Figure 5 illustrates an antibody against a mouse tumour (AKR-A) tested in a mouse model and demonstrates that, under appropriate conditions, you can get a significant cure rate for that cancer. Based on such studies in animals, experiments in patients with certain forms of leukaemia are now being carried out with toxin-loaded antibodies, and again the results look promising.

Figure 5

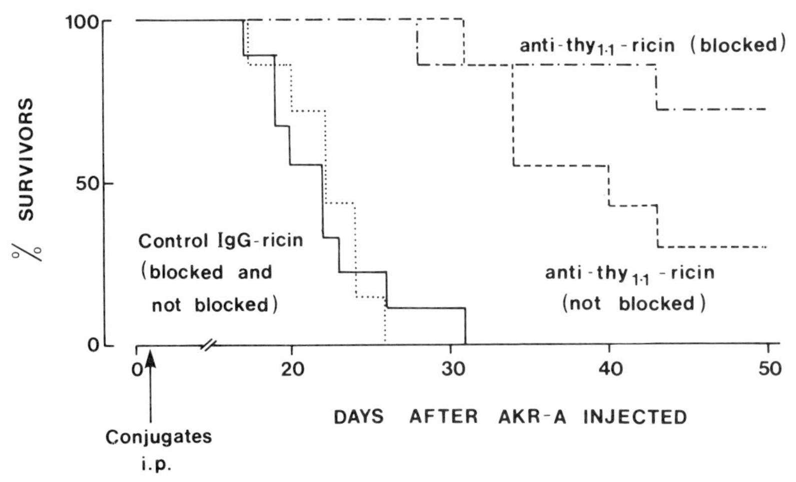

Anti-tumour action of anti-thy$_{1.1}$-ricin in a mouse model.

Interferons and related substances

The nude mouse model system provides a means of testing a variety of novel types of therapy. Some 10 or more years ago there was a great deal of publicity about interferons, the antiviral substances that the body produces when it is attacked by a virus. This is a very interesting set of proteins that have been shown to have some activity against cell growth. There are now some tumours that can be treated successfully with interferon, but it is a small number. Nevertheless, studies continue on interferon and indeed on a whole range of similar molecules called cytokines. These molecules are of great interest and some experiments, using them in combinations, suggest that they may be useful in direct therapy. Figure 6 shows some data from experiments with a human tumour model that in nude mice produces very rapid death. If tumour necrosis factor or interferon-γ are administered on their own there is a slight protective effect but, if they are administered together, there is a synergistic effect leading to a significant cure rate of perhaps up to 40% of tumours. This sort of experiment has encouraged clinical testing of novel combination therapies using cytokines, both with each other and with more conventional drugs.

Cancer spread or metastasis

One of the main features of the cancerous process is that, when a cancer starts in a particular site where the first genetic changes occur, it can be com-

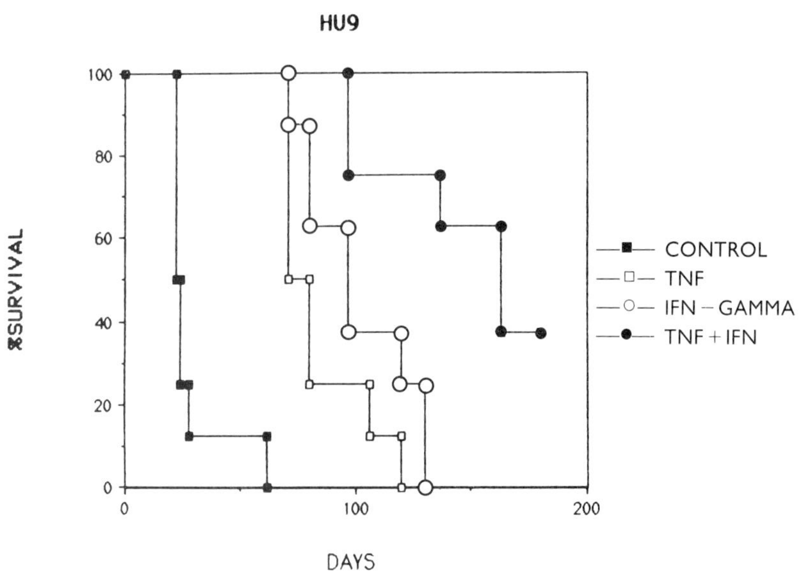

Protection, by tumour necrosis factor (TNF) and interferon (IFN) in combination, against a human tumour model (HU9) that produces rapid death in nude mice.

pletely cured by cutting the cancer out. The trouble is that a cancer can spread, and when it spreads and seeds itself in different parts of the body (i.e. when it 'metastasizes'), the disease is very hard to cure. In such cases, systemic therapy is needed; namely, the administration of drugs, or other forms of treatments, that will destroy the cancer cells wherever they are in the body. The process of metastasis is very complex. The tumour cells have to pass through various membranes and get into the lymphatics or the blood vessels for transporting them to other parts of the body. Then, the aberrant cells have to emerge from the vessels and seed in different places. Each of these steps must be understood since, if we can attack the process of metastasis, or if we can differentially attack metastatic cells, we again may have a novel treatment for cancer. But metastasis is essentially a process involving the whole organism and it is impossible to study it properly without carrying out those studies in the whole animal.

In such investigations it has been possible to select variants with different metastatic ability. For example, a human melanoma that can be grown in mice can produce both a low metastatic variant, that produces few metastases, and a high variant that produces many secondary tumours. A study of the properties of these variants can help us to understand the features that control metastasis. One of the features that has been shown to be important is the specificity of the attachment of tumour cells to the extra-cellular matrix (i.e. the material on which many cells lie). There is a very important family of molecules called integrins, which help cells attach to different components of this extracellular matrix. The receptor at the attachment site can be blocked by certain peptides which have a molecular configuration similar to some of the integrins. The same peptides will also block metastasis, illustrating the importance of the corresponding integrins in this process.

For some of the other integrins, however, the use of peptide analogues has shown that attachment of the integrin to the extracellular matrix receptor works in the reverse direction; i.e., that the attachment is actually required for the normal growth and differentiation of cells. In this case the ability to express attachment receptors is lost, since the cell wants to lose the ability to differentiate because this ability is not compatible with growth and tumour formation. These receptors are thus extremely important and again their functions can be studied by putting human tumours, modified in various ways with respect to the receptors, into suitable animal models and studying the effects on the process of metastasis. We now know, by staining tumours for the presence and absence of the receptors, that important changes in the expansion of receptors do take place during tumour development.

Transgenic mice

Transgenic mice are becoming a major tool in cancer research because they make it possible to combine our understanding of what goes on at the molecular and cellular levels with what happens in the organism as a whole. The many similarities between the genetics of the mouse and the human have made the mouse an important model organism of restudying human genetics. Indeed it is the main experimental mammalian organism for our work. The development of quite remarkable technology now makes it possible to take a cell from an early mouse embryo, grow it in tissue culture, introduce a mutation into a gene at will, and then reimplant that cell into another early embryo. Thus, you can obtain a mouse that carries the specific mutation you wish to study—a true form of genetic engineering. We anticipate using this technology to generate a mouse equivalent of the polyposis trait mentioned previously. Many specific changes in genes are already being studied using transgenic mice and I will conclude by giving one or two examples of the application of transgenic animals in the oncology area.

Firstly, it is possible to implant into mice the oncogenes that we know are mutated in human cancers. In this way, by implanting the genes that we know are changed in human tumours, it is possible to divert the spontaneous production of tumours in mice towards the production of the sort of tumours that occur in humans. Such tumours growing in a mouse should provide a good model system for many studies of human cancer, particularly for testing treatments that should work in humans.

Another example concerns a virus (originally discovered as a virus from milk) that can cause cancer in mice but only does so during lactation. The reason for this is that the virus can integrate next to a gene, called int-2, allowing it to activate that gene which then contributes to the development of cancer. The fact that the activation process only occurs during lactation is presumably due to hormonal effects. This process can be reconstructed in the mouse, with tumours developing only after several pregnancies. Thus, it is possible to recreate in some detail, in an appropriate animal model, the changes in particular tissues that we need to study and the sorts of tumours that may be found in different human situations.

Conclusions

The evidence I have presented illustrates that modern developments in technology have brought cancer research to a very exciting stage. In recent years our understanding of the processes underlying the development of cancer—from cell to organism—has increased enormously and we have been able to develop a number of new approaches to treatment. However, to achieve success in the ultimate treatment of cancer, the new technology must be implemented alongside animal experiments. Animal experiments only make up a very small part of the work that is carried out at the ICRF, but they are nevertheless very important. Most of our laboratories depend in one way or another on animal experiments, although their use of animals (mainly mice or rats) may be minimal. In fact, over the past 10 years, while the scientific staff at the ICRF has increased threefold, the directly attributable expenditure on animals has gone down by a factor of two. The trend in our work is towards using fewer animals, but to use the ones that we do in a more sophisticated and productive way compared with when the ICRF was founded in the early years of this century. However, despite all the recent advances in technology, I cannot envisage a time when the so-called alternatives, which have been so well described by David Hubel, could ever substitute for the essential experiments we carry out with animals. I cannot see a time when it would be possible to be at the forefront of cancer research, with all that we can do now, and not do animal experiments.

The moral and ethical issues arising from the use of animals in research have, of course, been much discussed. I think that the central question to be faced is whether the achievements of medical research in the past, and what is potentially possible in the future, justify the use of animals. For those of us in medical research who believe that the lives of people take precedence over those of other animals, the answer to this question must be clear. Those who do not believe this must surely have to tell us how many mice experimented on are—for them—worth a human life.

I believe it is extremely important, as has been emphasized several times in this book, that the scientists whose work requires them to use animals explain what they are doing and justify it to the public more openly and more actively than has been done in the past. The whole issue of public understanding of science, and the role of the scientific community in helping to create that understanding, is extremely important. In this connection there are many approaches that can be taken to increase the role of scientists. One that I believe may have some merit is in encouraging GPs to become involved. Similarly, I believe that some education in the imparting of scientific information to the lay public should be a part of all research training— perhaps everyone who takes a PhD in a scientific subject should, as a condition of that training, undertake some exercise in the presentation of science to the public. This is something that can and should be done, and the Royal Society is beginning to ask that its University Research Fellows do this. We have plans, through the Committee on the Public Understanding of Science (COPUS, a joint committee for the Royal Society, the British Association and the Royal Institution), to persuade the research councils and other organizations that this should be a requirement for obtaining a PhD.

I think that it is essential that we put out the correct facts about what is done and that we do not allow the sort of mis-statements, of which we have heard so much, to be widely promoted. It is hard to understand how anyone can make statements—of the sort disseminated to school children by the various antivivisectionist groups—claiming that animal experiments have never made a significant contribution to medical research, or that AIDS may have come out of a laboratory. These are statements that are patently untrue and one must conclude that either the people that make them know they are untrue and are deliberately spreading misinformation or that they are peddling their ignorance in a remarkable way behind an apparent scientific respectability. Whatever beliefs one holds on the moral issue, the scientific facts must not be distorted. That is where the scientific role is so important in presenting those facts dispassionately.

It must be a matter of enormous concern that it is children at school who have been selected as particular targets by the antivivisectionist groups. It is children who are being bombarded with half-truths and lies about how animal experiments cannot work. This seems to me to be an immoral stand to take and one that we must attack most vigorously. At the other end of the scale, the notion has been voiced recently that a reduction in the use of animals could be brought about by persuading people to leave their bodies for experimentation. This is really a most mischievous suggestion and one that could indeed militate against the valuable use of donor cards for making organs available for transplantation.

How can one have a medical organization, as was referred to by Colin Blakemore, that can actually declare itself in favour of abolishing vivisection, and claim that the majority of doctors now accept that experiments involving animals are of no practical value? Can a medical practitioner who believes that, in all honesty continue to be a practitioner? That seems to me to be impossible. I do believe that we must speak out most forcefully against these lies and that we must educate the scientific community to be much more active in this. If we are successful, then we will be able to take on board the fantastically exciting possibilities that confront us now in cancer research and in medical research as a whole—possibilities which will continue to give us extraordinary advances, but will continue to be dependent on animal experiments.

10

Final discussion

Chairman, David Jack (Research Defence Society): You have been presented with a number of different viewpoints, but the one thing that comes through very clearly is that experimental research in animals is an essential part of the process whereby we increase our understanding of mammalian biology. And it is, I think, useful sometimes to remind ourselves that we think with what we know. It is not always understood but that is so. And more importantly we think with what we know and understand. Now, animal experiments are an essential part of the acquisition of knowledge, as is work on subcellular fractions, cells, tissues, systems and so forth.

Equally, there can be no doubt that work on animals and on the other experimental approaches have led to amazing advances in the treatment and prevention of diseases. That is self-evident and you heard that from Professor Peters and from many others today. Looking forward, I think you have a similar message—that to deal with the huge problems that remain requires additional understanding, and thus continues to require animal work.

This last discussion will focus on the title of this book: 'Animal Experimentation and the Future of Medical Research'. Someone I believe said earlier that we are preaching to the converted and Sidney Brenner said: 'no, we are preaching to the preachers'; we hope that you will go from here as preachers.

Kurt Hellman (Westminster Hospital): Can I just make a practical point and that is that there are many patient associations who should be targeted in preference to audiences like this. As Professor Hubel has mentioned, if you send out literature to all of the doctors in the country, it is unlikely to get further than their desks. Patient associations, by comparison, would be much more inclined to listen and probably take some active steps.

Owen Wade (Birmingham University): I agree with the last suggestion. Certainly the association of patients with Parkinsons' Disease has been very receptive in discussions on this issue.

Can I make two points that I have found when I have been dealing with lay people over the problems of animal experimentation. One of the recurrent suggestions that is made to me is that I should stop experimenting on animals and start experimenting on prisoners—especially rapists. That is a theme that reflects an extraordinary attitude towards human beings.

The other point that is frequently raised is the issue of testing cosmetics. We all know that this is only a very minor part of animal experimentation, involving I think something of the order of 13 000 animals a year. Now, this causes great concern and anxiety amongst school children and others. I don't know whether any of the Home Office Inspectorate are here, but it would be a great help to people like me if they could explain briefly what work is currently being carried out on cosmetics. What little I know about cosmetics is that they are extremely bland and that the chance that there is any actual pain or discomfort produced in animals is absolutely minimal. The procedures must be very minor. Nevertheless this is something about which I am constantly questioned, and I just don't know the appropriate answers. Many 'cosmetics' are important medically because they may be applied regularly to sensitive skins for many years if not decades. Some of my patients need to apply them regularly for protection from u.v. light or for scar obliteration, and infants need protection from nappy rash.

David Jack: Testing of cosmetics has little to do with the future of medicine. Only about one half of 1% of the animals used are involved in cosmetic testing. Frankly it wouldn't worry me if no project licence was ever granted again strictly for establishing the safety of a cosmetic.

Sir Walter Bodmer: I do think it is very important to keep the issue of animal experimentation for medical research totally separate from that sort of toxicity testing, or other uses of animals such as factory farming and so on. If one does not, then one really gets into a confusing situation. I think it is extremely important that when we speak out for animal experimentation for medical research we do not confuse that with other issues upon which there may be very varying views on what is the appropriate attitude to adopt.

David Jack: I think new materials for cosmetics for the skin are very likely to come from dermatological research and, as such, the work will have a serious purpose.

John Subak-Sharpe (University of Glasgow): It has been suggested that we mention points that might help in emphasizing where animal experiments are needed. One aspect of research and development that has really not been emphasized today is the production of live vaccines. It is now very frequently possible, because we have already have the whole genome sequence of sometimes quite large viruses like the herpes viruses, to endeavour to identify and mutationally change those genes that are involved with pathogenicity, for example, neuropathogenicity. In fact this is the type of work in which we are presently engaged. You cannot possibly achieve this in tissue culture; you really require an animal model and in my view there is no way of side-stepping that. Moreover, if you look at the development of most viral vaccines in the past, they have been produced by an attenuation of strains of virus that were previously quite virulent, by means of multiple passages through individual animals. Most people are very happy to accept and make use of vaccines that will protect their children without recognizing how much animal experiments have contributed to the development of these in the past. This important aspect had not been emphasized.

David Jack: Thank you. There are of course many, many particular fields of research for which animals are absolutely essential.

Steven Rose (Open University): I have been reflecting during the discussions today and there is something that continues to trouble me. Within this room there is an air of calm rationality, in which we assert that what we are doing is right and that the problem is simply one of educating the lay public. The animal rights organizations have lots of money, lots of propaganda and lots of emotion. On the one hand we claim they are ignorant; on the other hand we have failed in an educational task.

Now I think that we have failed in an education task—I do not want to dispute that. But I do not think that it is entirely an educational task that we are dealing with, for the one issue that we have not resolved is that raised by David Hubel. Why is it that the animal rights movement is so powerful in England in particular and why is it so much more powerful now than say, 10, 20, 30 or 40 years ago? What is changing about our scientific situation and our sort of research *vis-à-vis* the rest of the world? What is changing in the perception, not just of animal rights activists, but of schoolchildren?

Despite all that COPUS, the Royal Society and educational packages can achieve, we are still going to be dealing with a very hard problem indeed.

Sir Walter Bodmer: I couldn't possibly argue against wanting to know more, but that does seem to me a rather negative approach to take. One cannot assume that things have necessarily got that much worse. The problems have always been there. In fact when the ICRF was first started there was so much antagonism to animal experimentation that the Prince of Wales at the time was very nearly prevented from becoming the President of the Fund. I think the degree of opposition to animal experimentation simply fluctuates. While of course one wants to know more about what underlies this, I cannot see that it cannot be right to get scientists to learn how to speak to the public about what they are doing, how it cannot be right to get more information into schools to counteract what the antivivisection groups are doing there, how it cannot be right, as Kurt Hellman has rightly said, to use patient groups to speak out, and how it cannot be right, as David Hubel suggested, to get the material through to doctors.

It is a broad, highly professional PR job that we are in need of and that is in fact a job which the research charity group that was mentioned earlier is planning to undertake (no doubt along with other groups). If you can carry out some good research to help this along, then surely so much the better. But we cannot wait for the results of that in order to do the things that must be done now.

David Jack: A number of distinguished speakers, each with a particular and different background, have explained how in their work there is some part for which animal experimentation is essential. This may be a major part as in pharmacology, or a smaller part as in some other disciplines; it varies from science to science. But be sure that it is quite impossible to mimic the sheer complexity of mammalian biology by any tests that are carried out in test tubes. Those who believe this to be possible have not understood the problem or the personal responsibility of those who introduce or approve the use of new medicines or procedures. We must not forget that the thalidomide disaster occurred because it was not tested in pregnant animals before coming into general use. Irrational constraints on animal experimentation will undoubtedly increase the risk of other serious adverse reactions and, perhaps even worse, impede the discovery of the better treatments needed by countless patients. In my view, our legislators, teachers and other opinion formers have a duty to ascertain the realities of animal-based research and be strong enough to resist well meaning but potentially disastrous changes in the law. We in biomedical science have a parallel duty to explain what we do and why.

Lord Adrian has already outlined the essential justification for animal experiments. I will not repeat it, for the principles he listed are enshrined in the 1986 Act. Our responsibility is to make sure that the Act is understood and that it is applied.

I would like to thank our speakers for addressing us and also those who participated in discussion.